Reducing Carbon Footprint in Different Sectors for Sustainability

Edited by Taha Selim Ustun

Published in London, United Kingdom

Reducing Carbon Footprint in Different Sectors for Sustainability
http://dx.doi.org/10.5772/intechopen.1001813
Edited by Taha Selim Ustun

Contributors
Abdullahi Mustapha, Betty Bastin, Bin Li, Chukwudi Emeribe, Dahl Winters, Emmanuel Ogbomida, Kamala Rajalekshmi, Kevin Fenton, Kwaku Boakye, Lawrence Ezemonye, Luca Caneparo, Marieke Versteijlen, Marleen Janssen Groesbeek, Mike Ajieh, Olurotimi Oguntola, Promise C. Anyanwu, Sandeep Sasidharan, Steve Simske, Taha Selim Ustun, Weihong Guo, Yuqing Zhang

First published in London, United Kingdom, 2024 by IntechOpen
IntechOpen is the global imprint of INTECHOPEN LIMITED, registered in England and Wales, registration number: 11086078, 167-169 Great Portland Street, London, W1W 5PF, United Kingdom

British Library Cataloguing-in-Publication Data
A catalogue record for this book is available from the British Library

Additional hard and PDF copies can be obtained from orders@intechopen.com

Reducing Carbon Footprint in Different Sectors for Sustainability
Edited by Taha Selim Ustun
p. cm.

This title is part of the Sustainable Development Book Series, Volume 22
Topic: Climate Change and Environmental Sustainability
Series Editor: Usha Iyer-Raniga
Topic Editor: Yixin Zhang

Print ISBN 978-0-85466-860-1
Online ISBN 978-0-85466-859-5
eBook (PDF) ISBN 978-0-85466-861-8
ISSN 2753-6580

For EU product safety concerns:
IN TECH d.o.o., Prolaz Marije Krucifikse Kozulić 3, 51000 Rijeka, Croatia,
info@intechopen.com or visit our website at intechopen.com.

IntechOpen Book Series

Sustainable Development

Volume 22

Aims and Scope of the Series

Transforming our World: the 2030 Agenda for Sustainable Development endorsed by United Nations and 193 Member States, came into effect on Jan 1, 2016, to guide decision making and actions to the year 2030 and beyond. Central to this Agenda are 17 Goals, 169 associated targets and over 230 indicators that are reviewed annually. The vision envisaged in the implementation of the SDGs is centered on the five Ps: People, Planet, Prosperity, Peace and Partnership. This call for renewed focused efforts ensure we have a safe and healthy planet for current and future generations.

This Series focuses on covering research and applied research involving the five Ps through the following topics:

1. Sustainable Economy and Fair Society that relates to SDG 1 on No Poverty, SDG 2 on Zero Hunger, SDG 8 on Decent Work and Economic Growth, SDG 10 on Reduced Inequalities, SDG 12 on Responsible Consumption and Production, and SDG 17 Partnership for the Goals

2. Health and Wellbeing focusing on SDG 3 on Good Health and Wellbeing and SDG 6 on Clean Water and Sanitation

3. Inclusivity and Social Equality involving SDG 4 on Quality Education, SDG 5 on Gender Equality, and SDG 16 on Peace, Justice and Strong Institutions

4. Climate Change and Environmental Sustainability comprising SDG 13 on Climate Action, SDG 14 on Life Below Water, and SDG 15 on Life on Land

5. Urban Planning and Environmental Management embracing SDG 7 on Affordable Clean Energy, SDG 9 on Industry, Innovation and Infrastructure, and SDG 11 on Sustainable Cities and Communities.

The series also seeks to support the use of cross cutting SDGs, as many of the goals listed above, targets and indicators are all interconnected to impact our lives and the decisions we make on a daily basis, making them impossible to tie to a single topic.

Meet the Series Editor

Usha Iyer-Raniga is a professor in the School of Property and Construction Management at RMIT University. Usha co-leads the One Planet Network's Sustainable Buildings and Construction Programme (SBC), a United Nations 10 Year Framework of Programmes on Sustainable Consumption and Production (UN 10FYP SCP) aligned with Sustainable Development Goal 12. The work also directly impacts SDG 11 on Sustainable Cities and Communities. She completed her undergraduate degree as an architect before obtaining her Masters degree from Canada and her Doctorate in Australia. Usha has been a keynote speaker as well as an invited speaker at national and international conferences, seminars and workshops. Her teaching experience includes teaching in Asian countries. She has advised Austrade, APEC, national, state and local governments. She serves as a reviewer and a member of the scientific committee for national and international refereed journals and refereed conferences. She is on the editorial board for refereed journals and has worked on Special Issues. Usha has served and continues to serve on the Boards of several not-for-profit organisations and she has also served as panel judge for a number of awards including the Premiers Sustainability Award in Victoria and the International Green Gown Awards. Usha has published over 100 publications, including research and consulting reports. Her publications cover a wide range of scientific and technical research publications that include edited books, book chapters, refereed journals, refereed conference papers and reports for local, state and federal government clients. She has also produced podcasts for various organisations and participated in media interviews. She has received state, national and international funding worth over USD $25 million. Usha has been awarded the Quarterly Franklin Membership by London Journals Press (UK). Her biography has been included in the Marquis Who's Who in the World® 2018, 2016 (33rd Edition), along with approximately 55,000 of the most accomplished men and women from around the world, including luminaries as U.N. Secretary-General Ban Ki-moon. In 2017, Usha was awarded the Marquis Who's Who Lifetime Achiever Award.

Meet the Volume Editor

Taha Selim Ustun received his Ph.D. in electrical engineering from Victoria University, Melbourne, VIC, Australia. He is currently a researcher with the Fukushima Renewable Energy Institute (FREA), National Institute of Advanced Industrial Science and Technology (AIST), where he leads the Smart Grid Cybersecurity Laboratory. Before that, he was a faculty member with the School of Electrical and Computer Engineering, Carnegie Mellon University, Pittsburgh, PA, USA. He has been invited to run specialist courses in Africa, India, and China. His research has attracted funding from prestigious programs in Japan, Australia, Europe, and North America. His current research interests include power systems protection, communication in power networks, distributed generation, microgrids, electric vehicle integration, and cybersecurity in smart grids.

Contents

Preface

In recent years, the imperative to reduce our global carbon footprint has become more urgent than ever. Climate change poses an existential threat, and addressing it requires a multifaceted approach that spans various sectors and disciplines. This book, *Reducing Carbon Footprint in Different Sectors for Sustainability*, is a comprehensive guide designed to provide insights and strategies across diverse fields, highlighting the critical efforts needed to mitigate environmental impacts and promote sustainable practices.

The energy sector, as one of the largest contributors to greenhouse gas emissions, is pivotal in the fight against climate change. The initial chapter delves into innovative approaches to reducing carbon emissions in energy production and consumption. These include advancements in renewable energy technologies, energy efficiency measures, and the transition from fossil fuels to cleaner alternatives. By exploring case studies and emerging technologies, a blueprint for a sustainable energy future is offered.

Transportation is another significant area of focus, particularly with the advent of electric vehicles (EVs). The chapter dedicated to this topic examines the rapid evolution of EV technology, infrastructure development, and policy measures that can accelerate the adoption of electric mobility. By reducing reliance on internal combustion engines, we can make substantial strides in lowering transportation-related emissions.

The cement industry, often overlooked in discussions about sustainability, is also a major source of carbon emissions due to its energy-intensive production processes. The book addresses innovative practices and materials that can reduce the carbon footprint of cement production. By highlighting cutting-edge research and practical applications, the relevant chapter aims to transform an industry that is fundamental to modern infrastructure.

Agriculture, another key sector, faces unique challenges and opportunities in reducing carbon emissions. The chapter on agriculture explores sustainable farming practices, carbon sequestration techniques, and the role of technology in enhancing agricultural efficiency. These insights are crucial for developing strategies that ensure food security while minimizing environmental impacts.

Architecture and urban planning play a significant role in shaping sustainable living environments. The book discusses the principles of green building design, the use of sustainable materials, and the importance of energy-efficient construction methods. By integrating these practices, architects and planners can create spaces that are not only environmentally friendly but also enhance the quality of life for their inhabitants.

Beyond these sectors, the roles of education and research centers are crucial in driving the transition to a low-carbon future. Dedicated chapters emphasize the importance of integrating sustainability into educational curricula and fostering a culture of environmental responsibility among future generations. Additionally, the book highlights the contributions of research institutions in developing new technologies and methodologies to combat climate change. By fostering collaboration between academia, industry, and policymakers, these chapters underscore the importance of a holistic approach to sustainability.

Reducing Carbon Footprint in Different Sectors for Sustainability is not merely an academic exercise but a call to action. Each chapter offers practical insights and solutions that can be implemented to drive real-world change. By bringing together expertise from various fields, this book aims to inspire readers to take proactive steps in their professional and personal lives to reduce their carbon footprint.

The journey toward a sustainable future is complex and requires concerted efforts from all sectors of society. This book serves as a valuable resource for policymakers, industry leaders, researchers, and educators who are committed to making a difference. Together, we can build a more sustainable world for future generations.

We hope that this volume will not only inform but also inspire action. The collective efforts of individuals and organizations across different sectors are essential in our global quest to mitigate climate change and ensure a sustainable future. It is our sincere wish that readers will find the insights and strategies presented in this book both enlightening and motivating.

Thank you for joining us on this critical journey toward sustainability.

Taha Selim Ustun
Fukushima Renewable Energy Institute,
National Institute of Advanced Industrial Science and Technology (AIST),
Tokyo, Japan

Dedication

For Hamza Suleyman, Halime Sedef and Merisa,
My constant sources of inspiration and strength.

Taha Selim Ustun

Section 1

Reducing Carboon Footprint of Different Sectors

Chapter 1

Carbon Footprint Reduction in Energy Sector with Higher Penetration of Renewable Energy Resources

Taha Selim Ustun

Abstract

Energy sector is a big source of carbon emissions. Traditionally, it is built, almost entirely, on fossil fuels with the exception of hydro and nuclear power plants. In line with the global mobilization toward tackling global warming by reducing carbon emissions, energy sector is trying to transition its generation portfolio to clean and sustainable energy sources. Renewable energy sources such as solar, wind, and wave provide energy without carbon emissions and provide a good alternative to traditional fossil fuels such as coal or natural gas. Another opportunity lies at the intersection of energy with transportation sector, another big contributor to carbon emissions. When electrified with renewables, transportation sector can both reduce its carbon emissions and provide support to electrical grid *via* novel solutions such as smart charging. However, these benefits come at a price. Non-dispatchable nature of renewable energy sources and their low inertia create operational and planning issues that make it hard to ensure demand is always met and the system operates in a healthy way. This chapter covers how renewable penetration can be increased while mitigating these issues with novel solutions. These include novel optimization ideas based on nature and new devices such as smart inverters.

Keywords: power engineering, renewable energy penetration, fossil fuel replacement, generation portfolio, smart grids, future grids, frequency control, power system optimization, nature inspired optimization techniques, electric vehicles, vehicle-to-grid (V2G), smart charging solutions, solar energy, wind energy, energy storage systems, sustainable energy, clean energy, power system inertia

1. Introduction

Industrial Revolution has surely transformed the humanity and its life on this planet. One of the biggest impacts of industrialization and burning more fossil fuels is the amount of carbon emissions and the resulting global warming [1]. The effects of this phenomenon can be observed in constant increase in temperature, melting icecaps, rising sea levels, and extreme weather events. Acknowledging the fact that greenhouse effects of carbon gases in the atmosphere cannot be ignored anymore,

 IntechOpen

nations around the world started carbon-reduction initiative such as Kyoto Protocol, EU 2020 vision, and so on [2].

Energy sector is one of the most prominent sources of carbon emissions as shown in **Figure 1** [3]. Traditional source of energy for electrical systems has been fossil fuels such as oil, natural gas, and coal. They have been a reliable bulk source of energy for large-scale power plants. On the other hand, their constant use made energy sector, and electrical systems, a big contributor to the global warming. For this reason, transformation of electrical system's generation portfolio has been at the center of carbon reduction and global warming mitigation efforts [4].

In principle, the idea is to deploy more renewable energy sources (RES) in the energy grid. The intended use might be direct electricity generation or an alternative approach such as direct heat capture. Some uses envision integration of RES at bulk generation stage, while some others are intended for much smaller applications at the distribution network level [5]. In addition to generation efforts, some attention has been diverted to increased use of storage devices and their alternative implementations such as electric vehicles being used as mobile storage devices. There are innovative solutions that use small-scale renewable energy resources to energize isolated communities that would, otherwise, be left without electricity [6]. These solutions focus on harvesting the available energy locally with minimal investment and provide energy to these marginalized communities at a low cost [7].

At every stage of such transformation, the power engineers face difficulties. This is due to the fact that traditional design of power networks was sketched thinking that energy flow will always be unilateral from bulk generation sites to bulk consumption sites that are clearly separated [8]. Although transmission lines have some flexibility to accept novel devices such as flexible AC transmission systems (FACTS), especially distribution networks are very vulnerable to such new deployments. They are not very stable; voltage and frequency values tend to experience big swings, and distribution network hardware such as the cables or transformers are bound by their lower electrical and thermal ratings [9].

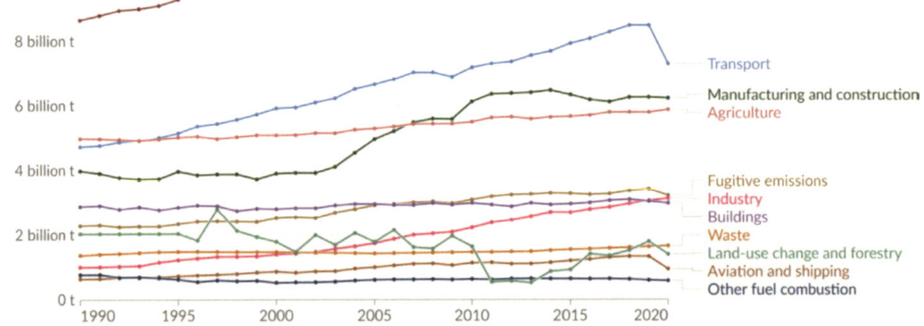

Figure 1.
Sector-based greenhouse gas emissions, CC BY [3].

In order to address these issues, researchers have been working on a myriad of ideas. The overarching question is *how to increase renewable energy penetration, and thereby decrease the fossil fuel share in power generation, while keeping the disturbances caused in system operation to a minimum?* Such solutions may be developed in the form of a novel device such as a smart inverter or virtual synchronous generator. Alternatively, a novel control approach can be developed that allows for more renewable energy and compensates for their negative impacts by being faster, more effective, and efficient. It is also possible to develop a holistic energy management system that combines novel hardware and software solutions in an integrated manner.

Rest of this chapter discusses the share of electrical power generation in global carbon emissions and the efforts toward increasing renewable energy share to reduce the overall carbon footprint of energy sector. It presents examples of novel solutions developed for various issues encountered in this process. Furthermore, it also discusses the side effects of these novel solutions, how they impact the overall system operation, and how they can be mitigated, again with examples from the literature. Finally, it draws the conclusions and provides perspectives for future research directions for interested individuals from the academia.

2. Energy-related global carbon emissions

Carbon emissions due to energy generation in all of its forms has been the leader in sectors related to greenhouse gas emissions. While the historical trend has been a steady increase in carbon release into the atmosphere since the early 1900s, a significant change in the rate of increase has been observed around 1950s. This trend has not changed ever since. It is true that the annual average rate of carbon emissions worldwide has been relatively lower in 2023, still with a 410 million tons of emission increase, and the overall emissions in 2023 reached 37.4 billion tons, a new record in history [10]. It is worthy to note that the increase in carbon emissions between 2019 and 2023 would have been much higher (around three times as much) had it not been for the positive impact of five key technologies for clean energy: electric vehicles, solar energy, nuclear energy, wind energy, and heat pump.

Figure 2 shows amounts and annual rate of change of carbon emission-related to energy sector in gigatons between 1900 and 2023.

For an optimist, there is good news in these numbers. They show that the clean energy initiatives around the world can and *do* make a difference. The carbon emissions growth in the past four years is kept to one-third of what it would have otherwise been. In this fashion, global economic growth and prosperity can be sustained with a very small change in the carbon emission growth. With the right amount of interventions and a bit of luck, that can even be kept to zero; that is, there will not be any more carbon emissions growth from year to year; instead, it will be kept constant.

On the other hand, a more cynical person, or a more *realistic person* as they like to call themselves more often than not, can interpret these results as all doom and gloom. Despite every country and sector doing their best to cut down carbon emissions, we could not prevent its growth. What is more worrying is that extreme weather events related to greenhouse effect, such as droughts, have had a strong impact on power generated from hydroelectric power plants and reduced the overall output from this, arguably only dispatchable, renewable source. Instead, fossil fuels had to be used that caused 170 million tons of emissions, more than 40% of the

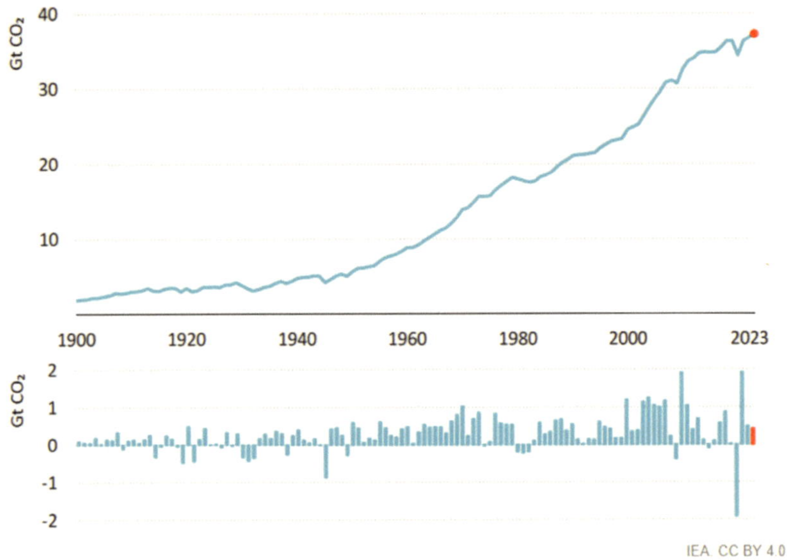

Figure 2.
Global carbon emissions related to energy, 1900–2023, IEA CC-BY [10].

overall carbon emissions growth last year. It is possible that the negative impact of global warming will trigger more extreme events that could reflect on these values more significantly.

A more neutral approach may help us understand that we are somewhere in between. We are nowhere near reaching goals set in Paris Agreement, far from it. However, we cannot overlook the fact that carbon emissions in advanced economies were reduced by around 4.5%, while GDP in these economies increased by 1.7%. Thanks to the transformations undertaken by the countries with advanced economies, their carbon emissions are back where they were roughly fifty years ago. The coal demand of these countries is the same as what it was in the 1900s. One of the reasons for this change is, unfortunately, due to the migration of industrial endeavors to other countries such as China where the annual emission growth was a whopping 565 million tons. Considering the global overall carbon emission growth was roughly 150 million tons less, 2023 would have been a year with negative carbon emission growth had it not been for China's carbon-intensive growth model. Carbon emissions per capita is 15% higher in China than in advanced economy countries. Global drought mentioned above caused roughly 33% of its emissions growth last year. India is another country that has been effected by the change in rainfall, where the weak monsoon season caused electricity production to move away from hydro. This resulted in a 25% increase in its emissions growth for the year 2023.

Figure 3 shows the total and per capita carbon emissions by different regions. It can be observed that European Union, Japan, and US have a downward trend in both the absolute amount and the per capita. China and India have both experienced significant carbon emissions, although India's annual growth is smaller.

To understand these trends, it is important to investigate the drivers behind these changes. As shown in **Figure 4**, the biggest change in the European Union was renewable electricity deployment. The second largest driver for this reduction in carbon emissions was lower industry-related emissions. Hydro and nuclear recovery, due to better precipitation and reactors that are put into service after much-needed

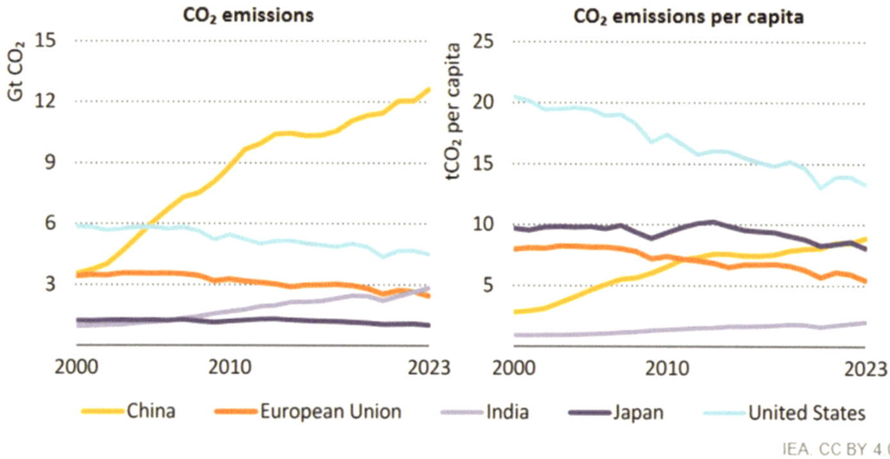

Figure 3.
Absolute amount and per capita carbon emissions [10].

maintenance, respectively, were significant factors as well. The biggest driver that caused an increase in carbon emissions was the GDP growth (**Figure 5**).

Lessons learned from the changes in carbon emissions of the US were significantly different. The biggest driver that caused the carbon emissions to fall despite a positive GDP change was the switch from coal to gas. Renewable energy deployments can be cited as the second biggest factor. However, that requires mentioning that half of this positive impact has been negated by the poor wind conditions and shortfall of the hydroelectric power generation. In other words, renewables had a positive impact, but their impact has been limited due to extreme weather conditions. This shows that a holistic approach is needed to reduce the carbon footprint of energy sector. Sometimes, switching from one fossil fuel to another may turn out to be the biggest positive impact.

A final look at **Figure 6** shows us that the biggest impact on the global carbon emissions is still due to the energy sector. Therefore, the potential for the biggest

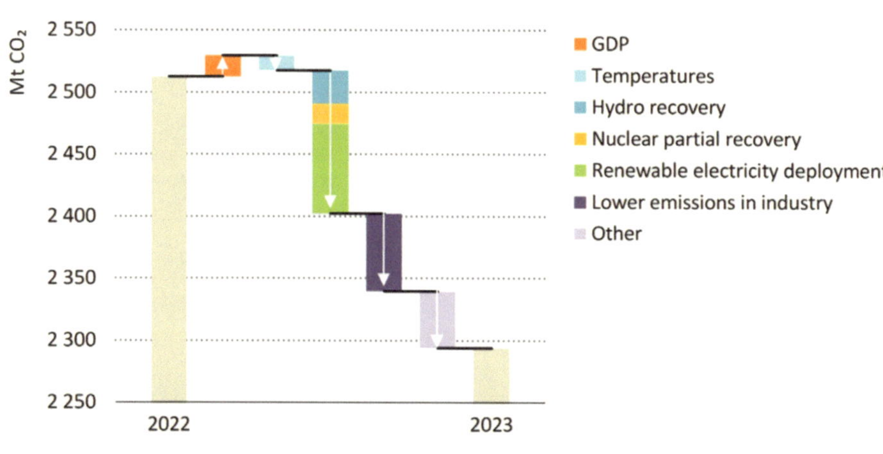

Figure 4.
Carbon emission change between 2022 and 2023 in EU [10].

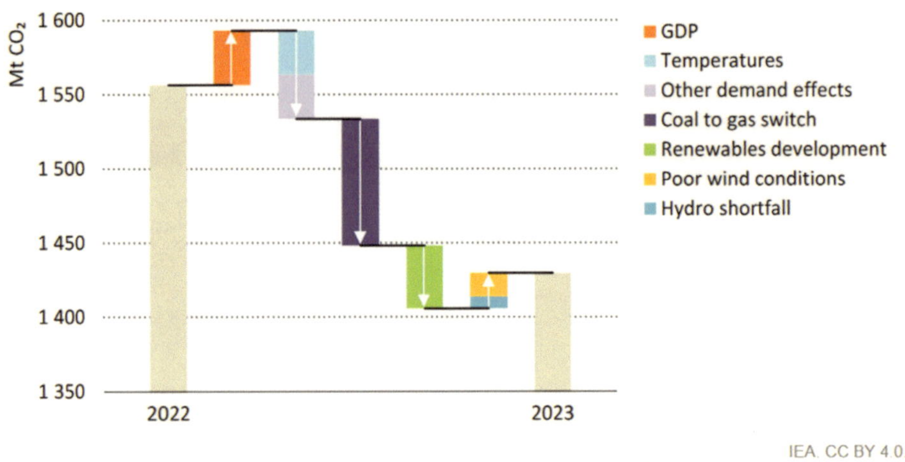

Figure 5.
Carbon emission change between 2022 and 2023 in the US [10].

positive impact or improvement lies therein. The second most influential sector is the transport sector. Clean energy use in transport is almost always associated with linking it with the electricity sector [12]. Electrification of transport sector, including ships, trains, and land vehicles, can be a game changer [13]. This requires proper planning and charging coordination with the electrical grid [14].

In all emerging market and developing countries (EMDE), the carbon emissions have risen for all sectors, that is, energy, transport, industry, and building sectors. For advanced economies (AE), all of these sectors have experienced a reduction. The world, in general, has experienced carbon gas emission reduction only in buildings sector. Considering their relative amounts, energy sector is the most influential and promises the biggest change.

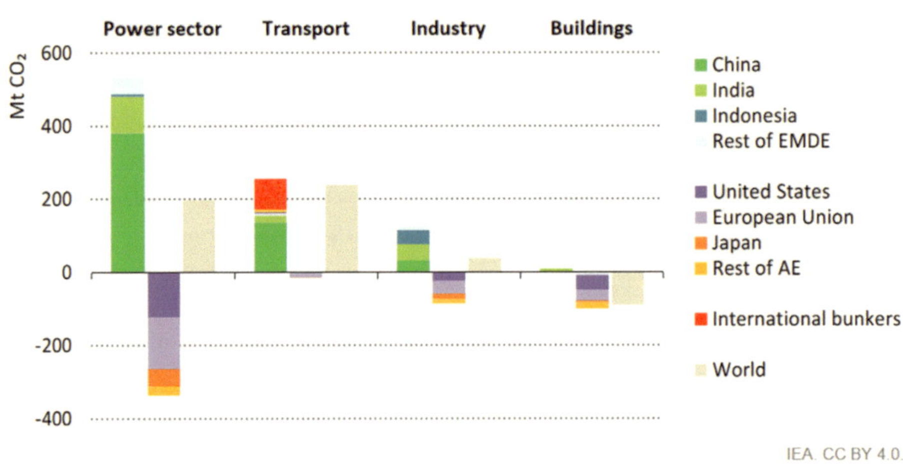

Figure 6.
Carbon emissions change by region and sector for 2023 [11].

3. Issues related to increasing renewable energy penetration in future grids

The previous section discussed what a big role energy sector in general, and electricity generation in particular, plays in global carbon emissions. Observing this fact, researchers have focused their efforts on transitioning from traditional fossil fuels to cleaner alternatives. Although it looks like a pretty straightforward decision, there are different barriers that hinder a mass migration to these alternative fuel sources.

First of all, these resources are intermittent, which means, unlike fossil fuels, they cannot be dispatched at will. This creates a massive issue for power system operation and planning [15]. From power system planning point of view, the golden rule is to always keep the demand and supply equal. This is not negotiable, and not doing so will have detrimental effects on the stability of the system that, in worst cases, may lead to blackouts [16].

In traditional power systems with fossil fuel fired power stations, it is relatively easy to manage this. Firstly, it is easier to estimate the consumption amount as the loads and consumer profiles are not as flexible and more predictable. Secondly, thanks to the storable nature of fossil fuels such as coal, oil, or gas, it is possible to have a large amount of power readily available and can be dispatched by increasing or decreasing the fuel input, thereby ensuring that power generation is equivalent to the consumption. Also, in the past, power networks used to have more operation margins that allow for changes. For instance, transmission lines were operating much lower than their thermal and electrical capacities. The only issue was finding the required energy and transmitting it.

Figure 7 shows a sample daily solar radiation in Sultanpur, India [17]. The profile is not smooth and has steep transients and constant variations. Furthermore, the solar radiation estimation calculations yield only the daily average solar radiation, not how it may fluctuate in time. Nor do they take into account sudden changes such as cloud coverage. Therefore, solar energy cannot be the primary source of energy in

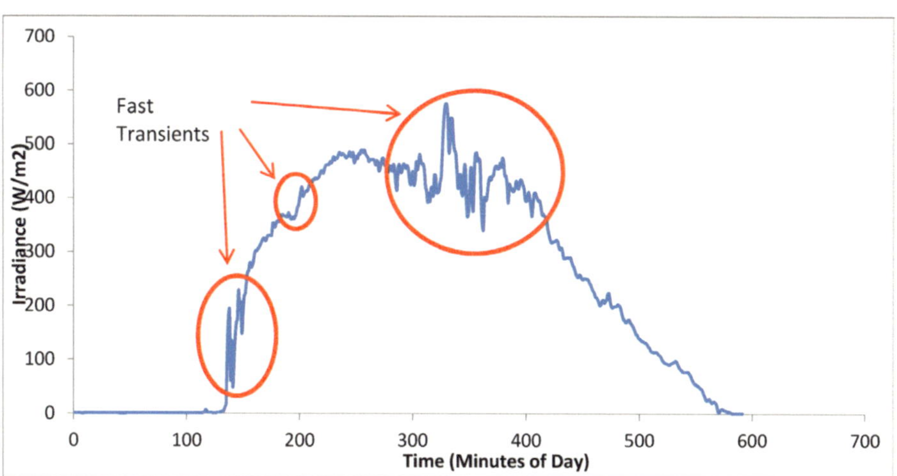

Figure 7.
Variations and fast transients in a typical daily solar radiation profile [17].

these systems. There should always be some additional measures, such as storage or dispatchable generation source, that can step in to meet the demand if solar energy fluctuates or diminishes. It is also important to note that most of the renewable energy sources are affected by the weather and seasons. Their energy output varies, and their seasonal variability needs to be compensated by other means. As shown in **Figure 8**, different seasons such as a dry season or a tropical one have significantly different energy outputs. For instance, PV output is very limited in a continental warm summer setting, while it can meet half of the demand in a tropical setting. It goes without saying that successful integration of renewables requires effective mitigation of both short-term and long-term variability in their nature [11].

As mentioned earlier, another major issue related to integration of renewable energy resources is in power system operation. This manifests itself in the stability and health of two key operational parameters, voltage and frequency. For example, **Figure 9** shows the voltage profile of a 100 V distribution network. On the bottom left, a network with no PV penetration is shown. Voltage gradually drops as one gets further away from the local transformer [18].

Bottom middle shows when PV panels are installed on these households, and they start injecting power into the grid. Since this is not what the distribution networks were initially designed for, a significant voltage rise appears. In some cases, this voltage rise goes beyond the higher limit. In order to avoid this, and although it means rejecting free and clean energy, the PV generation is limited, that is, curtailment.

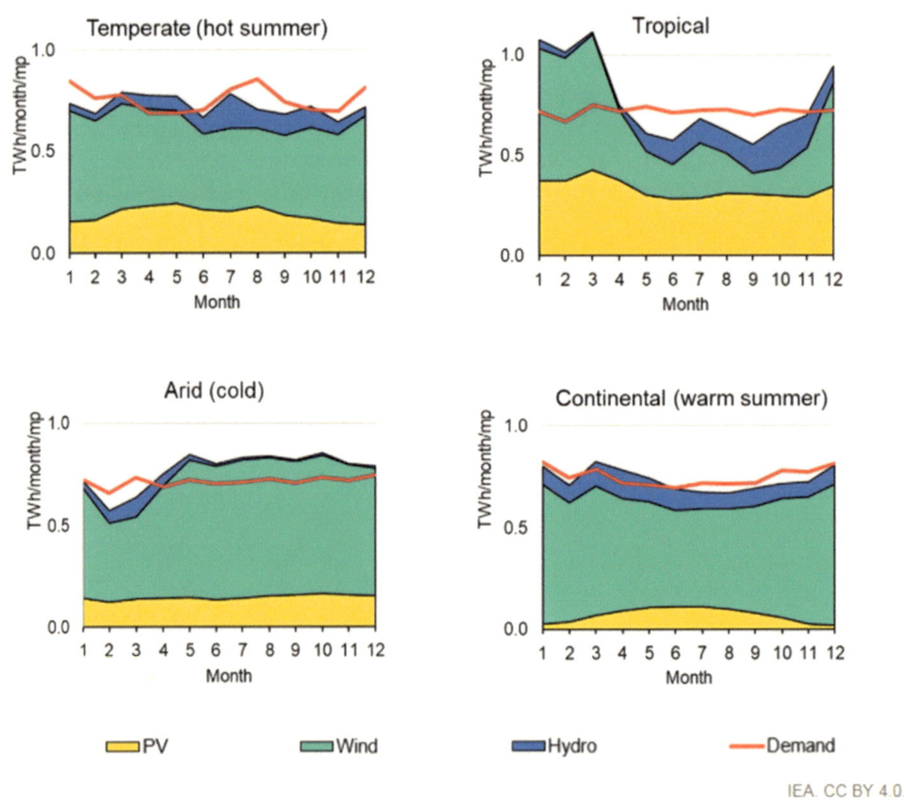

Figure 8.
Seasonal variations in the energy yield of renewable energy resources [11].

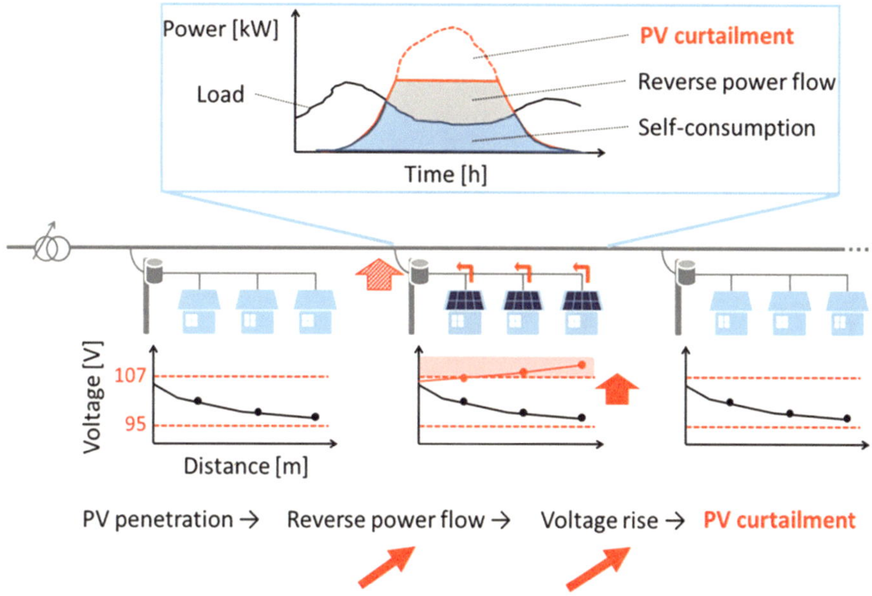

Figure 9.
Voltage issues related to local renewable energy generation.

This shows that due to physical restrictions, it is not possible to harvest all the available renewable energy at a particular location. Even if, feasibility studies show a much higher harvestable potential. In theory, if all the transmission and distribution networks are fully refurbished, they can accommodate more renewable penetration. However, it is not realistic to expect this due to high costs, regulatory limitations, and public disapproval.

Issues related to the stability of the frequency are caused by the inverter topologies that are used to interface most of the renewable energy resources and storage devices to the existing grid [19]. These topologies are built by solid-state electronics that switch at a very high frequency to mimic AC systems. Traditionally, the power grid's frequency is set by rotational motion that takes place inside the synchronous generators. Due to their physical structure and rotor rotating inside an electromagnetic field, these systems have an inertia that resists sudden changes in the interconnected system. This is an excellent security mechanism as it inherently gives frequency stability to the system. As shown in **Figure 10**, inverters have several legs of solid-state inverters, some filters, and advanced measurement and control loops [20]. They lack any rotating parts that would give an inertia to their operation. This is not a big issue if the penetration of renewable energy, and the number of inverters that are used to interface them to the grid, is not high. It becomes an issue when they are used in very high penetrations, in isolated areas or weak grid situations, for instance, places that are very far away from the grid to benefit from its stability.

Finally, it is possible to cite lack of necessary market and economic tools as a barrier as well. Researchers have focused on mitigating technical and physical limitations that hinder deep penetration of renewables. Recently, it has been acknowledged that even if technical barriers are entirely removed, there are regulatory and financial barriers in place [21]. This includes centralized bidding structures where small-scale or individual power producers cannot participate, regulatory limits about how to sell

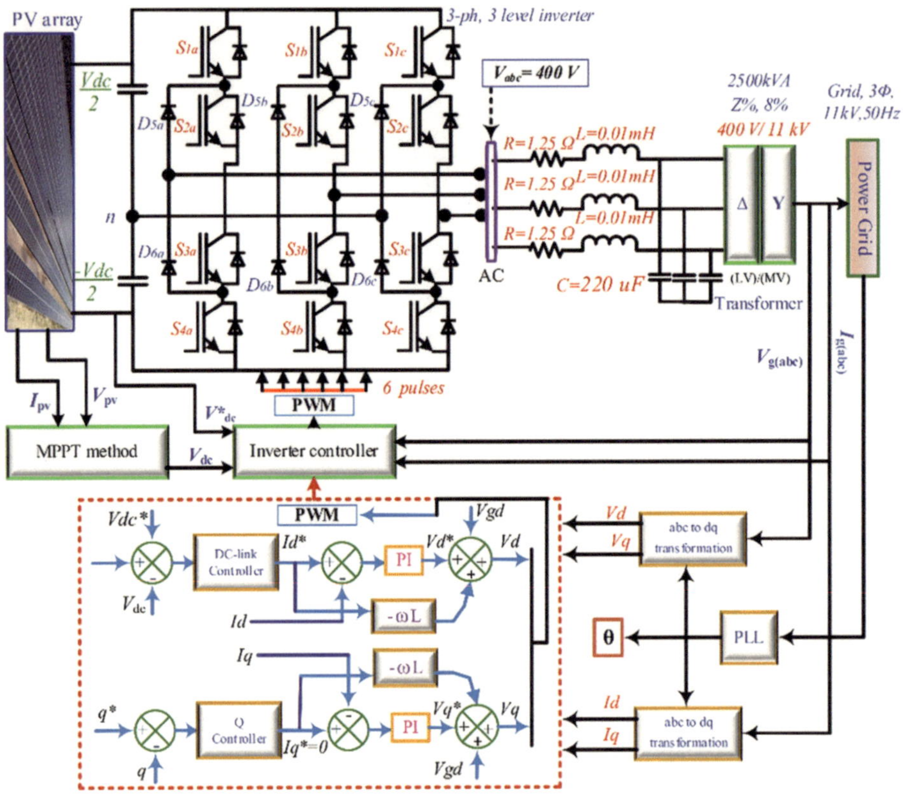

Figure 10.
A typical inverter topology with advanced control [20].

energy and which channels to use, bureaucracy around obtaining necessary licenses as a producer, consumer, or prosumer etc. Assuming that almost every household or electric vehicle owner can participate in the market, the number of participants and the information technology infrastructure required to deal with an energy market with that many stakeholders make it very challenging. Traditional energy markets with centralized market operators and only a handful of participants cannot meet the needs of a power grid with very dynamic and active market.

4. Solutions toward mitigating issues related to higher penetration of renewables

Researchers have investigated different approaches to mitigate various issues mentioned in the previous section. When it comes to managing non-dispatchable behavior of renewables, the solutions focus on creating a generation mix, which may or may not include storage systems, to address intermittent and unpredictable nature of renewable energy resources. The very basic approach towards this is ensuring that the capacity of the storage systems or dispatchable generators is equal to the critical loads in a system. In other words, even in a contingency case where the renewable energy resource is completely lost, due to cloud coverage for PVs or storm conditions for wind turbines, the system can still sustain itself. An example microgrid that is designed in a fail-safe manner is shown in **Figure 11** [22].

DOI: http://dx.doi.org/10.5772/intechopen.1005769

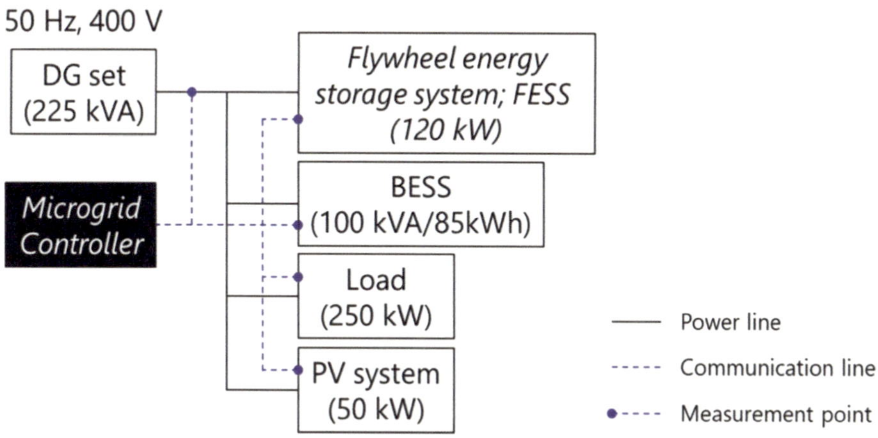

Figure 11.
Microgrid with renewable energy sources and storage system [22].

It is also possible to use grid-connection as a fallback mechanism, in case the renewable energy sources fail to generate the estimated energy as shown in **Figure 12**. Such systems can be designed to harvest as much clean energy as possible, when available. In other cases, grid connection can always be utilized to make up for the demand-supply mismatch.

In both cases presented above, the main objective is to build the most optimal system so that the available renewable energy is harvested as much as possible. Some researchers have attempted to combine different storage technologies to come up with a better cost-performance ratio and a longer life cycle. One such attempt is shown in **Figure 13** where battery and supercapacitor pair is utilized to make use of different characteristics of these technologies [24]. Batteries need longer times to charge, have a limited discharge power, and are affected by charge-discharge cycles which reduces their lifetimes. However, they can come in large capacities and are relatively cheaper. Supercapacitors, on the other hand, can charge and discharge very quickly, are expensive, and have better charging performance.

The benefit of such solutions is that if the intermittency of renewable energy, in this case solar, is managed more effectively, it is possible to rely on it more and increase its share in the generation portfolio. In case of any issues, the balancing system will compensate and ensure reliable operation of the system. In the long run, more renewable energy is capture and fed into the system. Especially for setups where

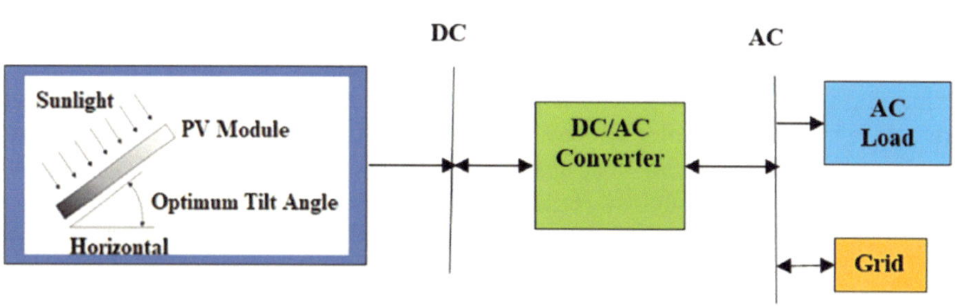

Figure 12.
Harvesting renewable energy sources with grid connection [23].

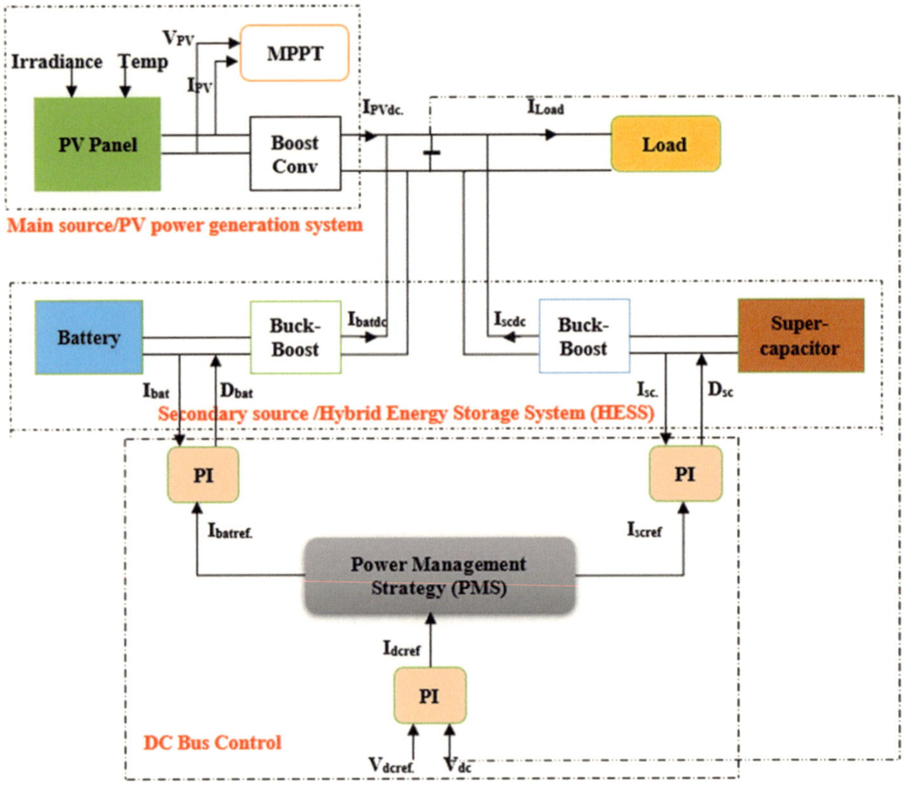

Figure 13.
Battery-Supercapacitor pair for enhanced solar energy capture [24].

fossil-fuel fired generators are used as fail-safe solutions, as an example see **Figure 14**, this reduces carbon emissions due to the power system generation [25].

A more integrated way of doing this involves communication with all system components, for instance, generators, loads and storage devices and, in some cases, weather stations for renewable generation prediction. These are called energy management systems (EMS) and are designed to monitor the condition of the power system and make decisions in real time. As shown in **Figure 15**, EMS controller maintains communication with all components and ensures safe and reliable operation of the power system while maximizing renewable energy capture such as from wind turbine (WTG) or central receiver solar thermal system (CRSTS) [26].

These decisions can be done with a simple flowchart, if-else mechanism, or lookup table. However, system based on such deterministic solutions fall short in maximizing renewable energy capture. Therefore, more advanced EMS approaches incorporate machine learning approaches such as artificial neural networks [27] or fuzzy logic [28] to search the most optimal operating point in the solution space. As shown in **Figure 16** [29], such systems may utilize more than one artificial intelligence approach [30–33]. The ultimate objective stays the same: maximizing the renewable energy utilization rate while managing intermittency to ensure power system operation stability.

One important aspect of these solutions is that they require extensive communication between all components present in a power system [34]. Designing a system

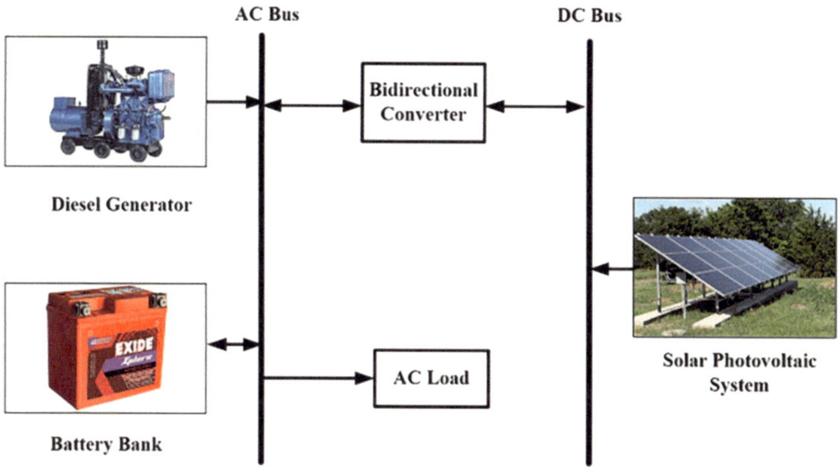

Figure 14.
A microgrid where diesel generator is used as a fallback solution when solar energy is not available [25].

where a macro player such as a power producer and a micro player such as an electric vehicle in a residential household can exchange mutually legible messages is a tall task [19]. This is especially so in power systems where extensive communication or message exchanging has not been traditionally implemented [35]. This requires a standard communication approach to be developed in power systems. While this topic is very broad and cannot be fully discussed in this chapter, interested readers are encouraged to refer to these works for further details [36, 37]. An unwanted result of such standardization is cybersecurity vulnerabilities that created unprecedented issues in power systems [35–43]. This is also a very recent and fertile research field that cannot be properly discussed here and needs its own discussion avenue. Interested readers can refer to these works as a start [44–50].

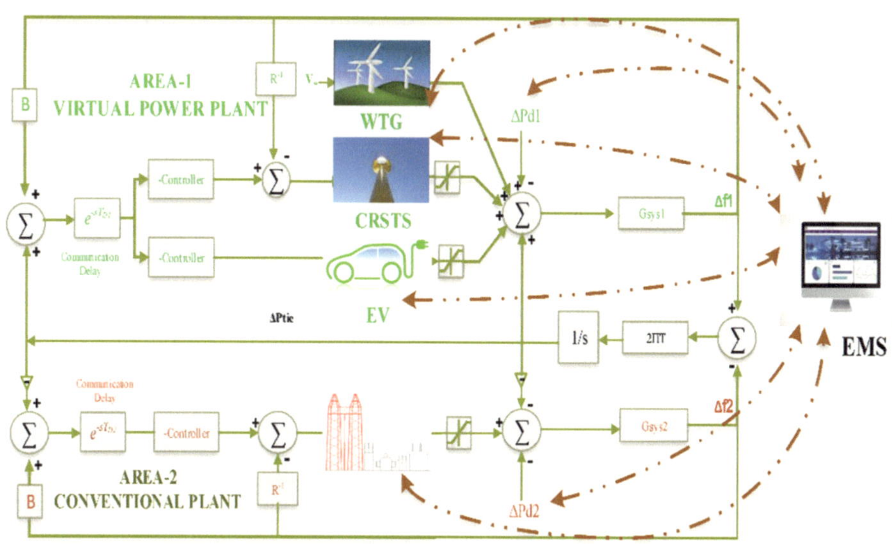

Figure 15.
EMS design to maximize renewable energy capture [26].

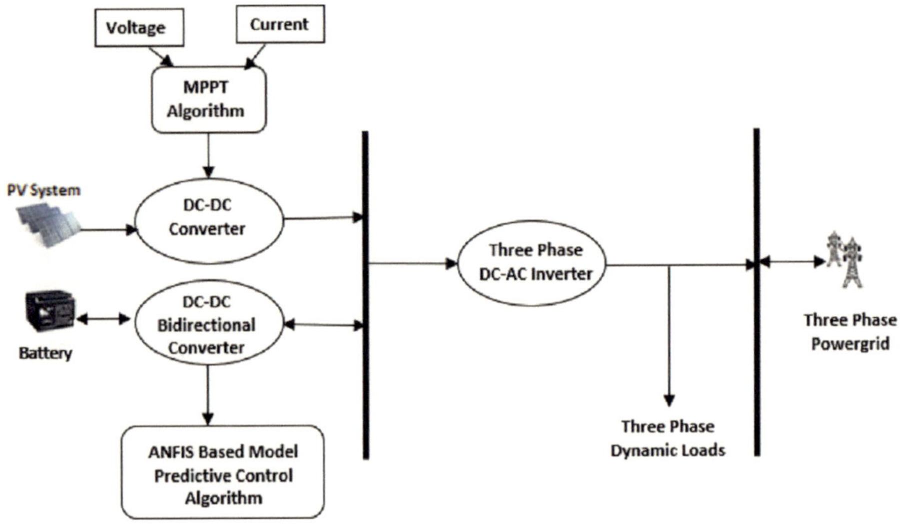

Figure 16.
Use of artificial intelligence in energy management systems [29].

The above solutions focus only on making sure that the demand and supply are equal in a simplistic manner. They do not take operation of power systems and system dynamics into account. This is important because when a generation source is lost, this has negative impacts on the voltage and frequency of the system. On the other hand, too much generation or active power injection causes these parameters to increase and reach their limits [37]. Both of these cases need to be carefully avoided. **Figure 17** shows how an increase in the load disturbs the system frequency and how its magnitude as well as its duration are dependent on the controller topology that is employed [51]. This shows that using more sophisticated controllers such as (IPD-(1 + I)) is advantageous. Some researchers focused on using cascaded controllers so that the system can be more robust and effected less by the disturbances. As shown in **Figure 18** [52], such topologies can prove to be much more stable and ride through disturbances with less impact and in shorter time.

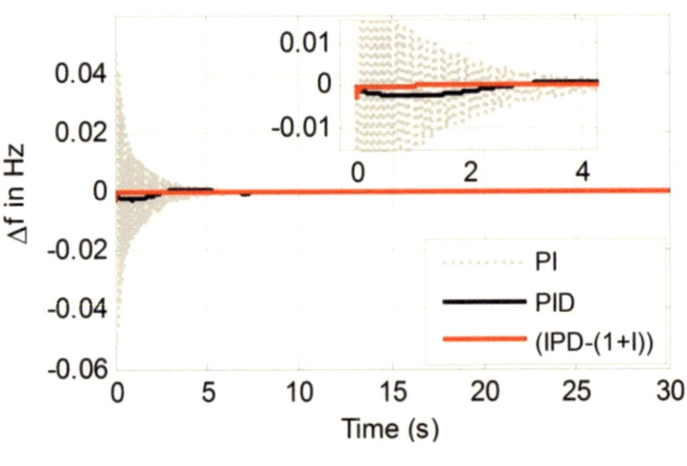

Figure 17.
Traditional controllers and their reaction to load increase in a system observed as a frequency disturbance [51].

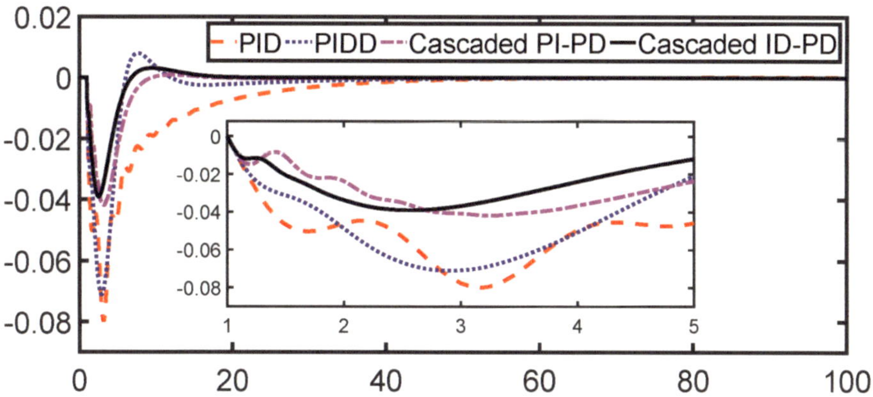

Figure 18.
Comparison of different controller topologies, cascaded controllers are more stable with less overshoots [52].

Figure 19 shows how a 1% increase in the system load disturbs the system frequency and how long different controllers need to stabilize it [53]. Here, two different optimization algorithms have been utilized to enhance the response time and operation of the controllers: African vulture optimization algorithm (AVOA) and a version of AVOA that is enhanced by sine cosine approach, that is, Sine Cosine-Adopted African Vulture Optimization Algorithm (SCaAVOA).

It is a very dynamic and active research field where such different optimization algorithms are implemented and benchmarked against each other under different power system conditions [54]. There are different inspirations for such optimization techniques. A recent development has showed that imitating behaviors observed in the nature may prove to be very useful in such situations [55]. **Figure 20** [56] shows how a power system can become more robust and stable when different nature-inspired optimization algorithms are employed. Expectedly, the more robust a system is, the more renewable energy deployments it can accommodate as it can effectively

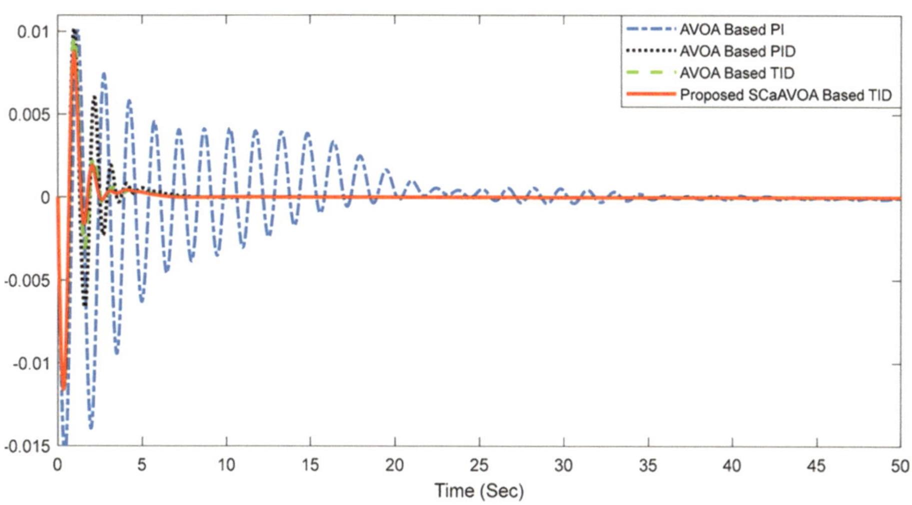

Figure 19.
Frequency disturbance caused by 1% increase in system load and its reduction with different control approaches [53].

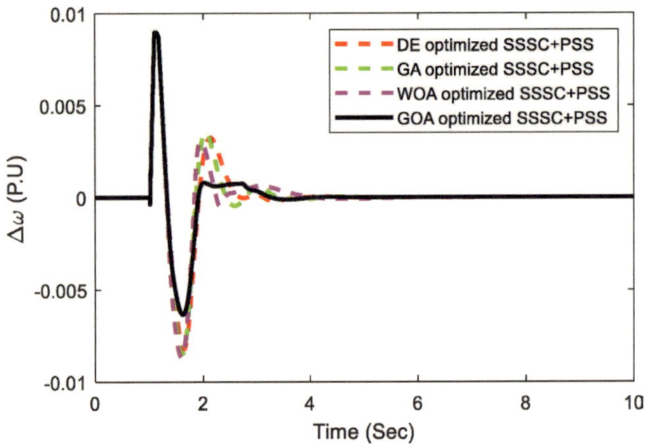

Figure 20.
Benchmarking of different optimization techniques to stabilize speed, differential evolution (DE), genetic algorithm (GA), whale optimization algorithm (WOA), and grasshopper optimization algorithm (GOA) [56].

mitigate the intermittent nature of these resources. It is very exciting to see that carbon footprint of power systems can be reduced with solutions received from the nature. In a strange way, nature helps lift off burden of carbon emissions from the nature itself.

In addition to controller and optimization approaches discussed above, there are full-fledged novel technologies or devices that are developed to support this transition. The two leading concepts are smart inverters and inverters with virtual inertia. Smart inverters, also known as inverters with advanced capabilities, are devices that can provide additional grid support such as reactive power injection [57]. **Figure 21** shows how smart inverters provide positive or negative reactive power into the system to provide stability, depending on the system voltage [58]. When this figure is contrasted with the case depicted in **Figure 9**, the distinction between conventional inverters and smart inverters becomes clearer. Due to their benefits, smart inverters are required by several grid operators for renewable energy deployment [59]. Similar implementation can be done to provide frequency support *via* frequency-watt control in smart inverters [60].

In a similar fashion, virtual inertia is an effort to mitigate the negative impacts of inverter-interfaced renewable energy resources on the system. Control mechanisms in inverters is revised to include resistance to voltage and frequency changes as a real electric machine with rotating parts would [61]. With this much needed change, the inverters are being converted from being grid-following into grid-forming inverters. When frequency-watt capability is combined with df/dt capability in smart inverters, virtual inertia can be successfully implemented. **Figure 22** shows control hardware in the loop testing for a smart inverter controller that has fully functioning virtual inertia capability [62]. Disturbances in the system such as load or generation change triggers a frequency disturbance. In order to recover frequency to its nominal value, that is, keep the system stable, smart inverter with virtual inertia capability exchanges reactive and active power to ensure that frequency change is limited and returned to its value.

The solutions and technologies discussed above focus on one thing and one thing only: to reduce negative impacts of renewable energy sources on the power system operation with an aim to maximize their penetration. This is the gateway towards

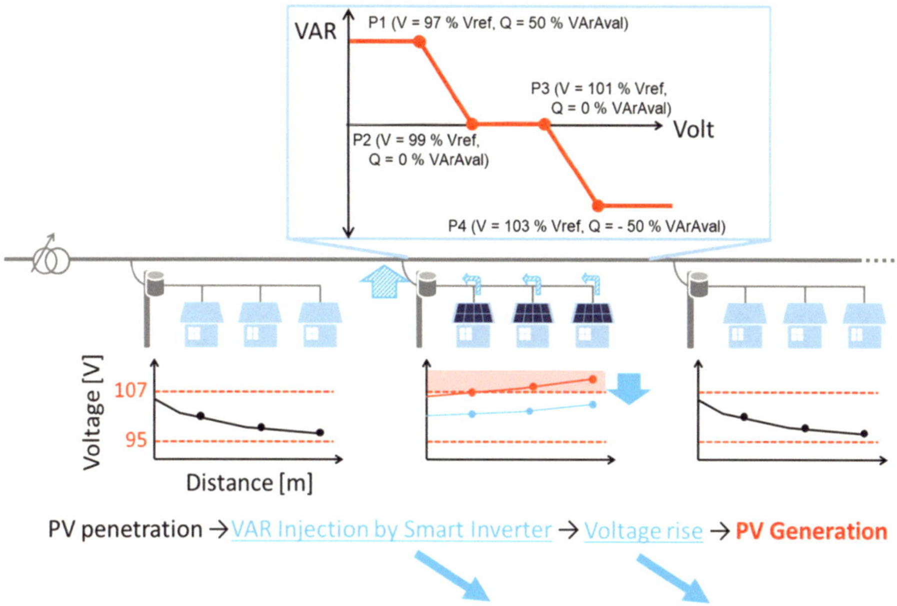

Figure 21.
Volt-Var functionality implemented in smart inverters [58].

reducing the share of fossil-fuels in energy sector and reducing the carbon footprint in a significant way. The final piece of the puzzle lies with the energy markets. Traditional markets were designed for bulk energy trading, which addresses mostly fossil-fueled power stations, with the exception of hydroelectric and nuclear power plants [63]. Successful integration of renewable energy resources requires these paradigms to change [64, 65]. Generation at much smaller scale should be possible; non-dispatchable generation should be accepted, and auxiliary services to stabilize the system should also be incentivized and monetized [66, 67]. Distributed nature

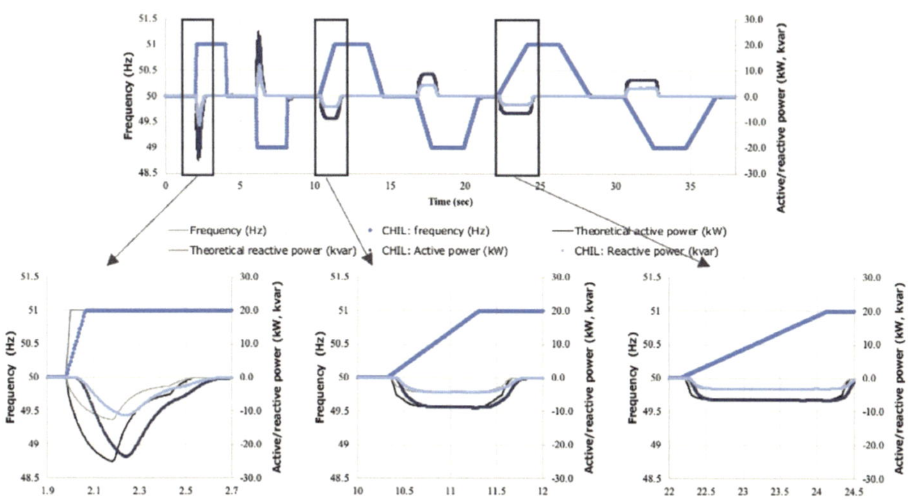

Figure 22.
Virtual inertia implemented in smart inverters [62].

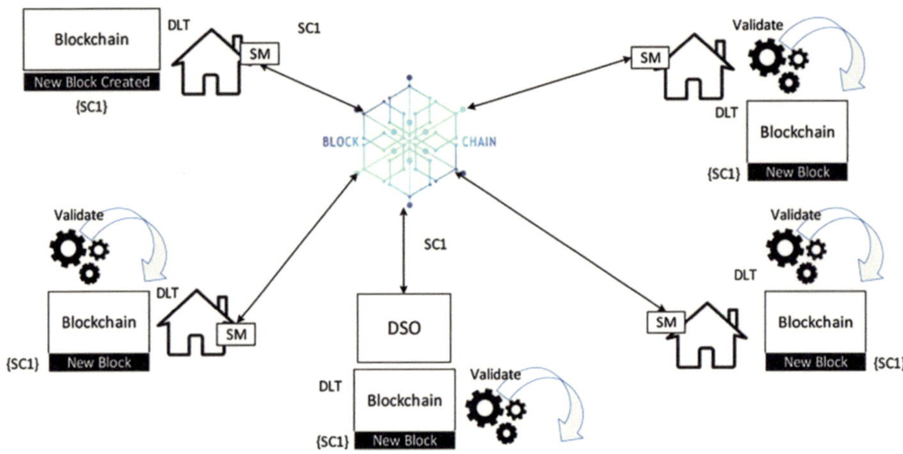

Figure 23.
Renewable energy trading with Blockchain technology [68].

of renewable energy resources and the fact that peer-to-peer energy trading is much more convenient lends itself to novel trading ideas such as Blockchain as shown in **Figure 23** [68]. This will unlock the potential of independent, distributed, and deregulated energy markets [69].

5. Conclusions

Replacing fossil fuels with clean renewable energy resources is not an option anymore. Changing weather conditions, more frequent extreme weather events, and their apparent impacts on life make it clear that a transition must be made. As one of the biggest sources of carbon emissions, power sector's transformation will have large impact. Furthermore, when integrated with the transportation sector, another big carbon emission source, an interdisciplinary solution can be developed.

The solution lies in harvesting clean and mostly nondepletable renewable energy resources and replacing fossil fuels with them. There is more than enough potential in alternative energy sources such as solar, wind, wave, and geothermal. However, their non-dispatchable and non-predictable nature creates issues for power engineers. It is not possible to fully rely on these resources without compromising the health and stability of the entire power system. Researchers have been working towards developing novel approaches, solutions, and devices that will mitigate these negative impacts. These solutions have been covered in this chapter in detail, what part of the problem they address, and how they do it.

This research field is very promising, and there are exciting opportunities for future work. Development of an inverter with a comprehensive virtual inertia functionality is an important step towards reducing carbon footprint of the energy sector. Integration of all power system systems and devices with a common language for better monitoring and control is an active research field. Due to competing interests, different stakeholders lock horns and make it very difficult to come up with an ultimate solution. Integration of household items and smart meters with larger systems creates opportunities for solutions such as demand side management and energy dispatch at the consumer end. However, these systems create cybersecurity and privacy issues that need to be tackled before such systems can be implemented at large scale.

Conflict of interest

The authors declare no conflict of interest.

Author details

Taha Selim Ustun
Fukushima Renewable Energy Institute, National Institute of Advanced Industrial Science and Technology (AIST), Japan

*Address all correspondence to: selim.ustun@aist.go.jp

References

[1] World Economic Forum. Here's How CO2 Emissions Have Changed Since 1900. 2022. Available from: https://tinyurl.com/56jr7dse [Accessed: March 11, 2024]

[2] United Nations, Climate Change. What Is the Kyoto Protocol? Available from: https://unfccc.int/kyoto_protocol [Accessed: March 11, 2024]

[3] Our World in Data. Emissions by Sector: Where Do Greenhouse Gases Come From? 2024. Available from: https://tinyurl.com/4tz8d5yh [Accessed: March 11, 2024]

[4] Our World in Data. Electricity Generation from Fossil Fuels. 2023. Available from: https://tinyurl.com/47u9xrun [Accessed: March 11, 2024]

[5] IEA. Renewables 2023. Paris. Available from: https://www.iea.org/reports/renewables-2023: IEA; 2024. Licence: CC BY 4.0 [Accessed: March 11, 2024]

[6] Hubble AH. Scaling renewable energy based microgrids in underserved communities: Latin America, South Asia, and sub-Saharan Africa. In: 2016 IEEE PES PowerAfrica; Livingstone, Zambia: IEEE; 2016. pp. 134-138

[7] Buchana P. The role of microgrids & renewable energy in addressing sub-Saharan Africa's current and future energy needs. In: IREC2015 the Sixth International Renewable Energy Congress; Sousse, Tunisia. 2015. pp. 1-6. Available from: https://ieeexplore.ieee.org/document/7110977

[8] Saadar H. Power Systems Analysis. 2nd ed. McGraw-Hill; 2004. Available from: https://www.amazon.com/Power-Systems-Analysis-2nd-International/dp/0071281843. ISBN 0071281843

[9] Federal Energy Regulatory Commission. Managing Transmission Line Ratings, AD19-15-000, Report. Washington DC; 2019

[10] IEA. CO2 Emissions in 2023. Paris. Available from: https://www.iea.org/reports/co2-emissions-in-2023: IEA; 2024. Licence: CC BY 4.0 [Accessed: March 11, 2024]

[11] IEA. Managing Seasonal and Interannual Variability of Renewables. Paris. Available from: https://tinyurl.com/yd9b8szy: IEA; 2023. Licence: CC BY 4.0 [Accessed: March 11, 2024]

[12] World Economic Forum. The Electrification of Transport Could Transform our Future—If We Are Prepared for It. 2018. Available from: https://tinyurl.com/y9wuss9y [Accessed: March 11, 2024]

[13] Department of Energy, Office of Energy Efficiency and Renewable Energy. Electrifying Transportation to Benefit every American. 2022. Available from: https://tinyurl.com/yckfws77 [Accessed: March 11, 2024]

[14] Nsonga P et al. Performance evaluation of electric vehicle ad-hoc network technologies for charging management. In: 2017 IEEE PES Asia-Pacific Power and Energy Engineering Conference (APPEEC), Bangalore, India. 2017. pp. 1-5. Available from: https://ieeexplore.ieee.org/document/8308933

[15] Climate Portal. Massachusetts Institute of Technology, Intermittent Versus Dispatchable Power Sources. 2021. Available from: https://tinyurl.com/5n9y8c7y [Accessed: March 11, 2024]

[16] Business Insider. Get Ready: More Blackouts Are Coming. 2023. Available from: https://tinyurl.com/2vpa49um [Accessed: March 11, 2024]

[17] Javed K, Ashfaq H, Singh R. Design and performance analysis of a stand-alone PV system with hybrid energy storage for rural India. Electronics. 2019;**8**:952

[18] Ustun TS, Aoto Y, Hashimoto J, Otani K. Optimal PV-INV capacity ratio for residential smart inverters operating under different control modes. IEEE Access. 2020;**8**:116078-116089

[19] Ustun TS, Ozansoy C, Zayegh A. Simulation of communication infrastructure of a centralized microgrid protection system based on IEC 61850-7-420. In: 2012 IEEE Third International Conference on Smart Grid Communications (SmartGridComm); Tainan, Taiwan. 2012. pp. 492-497. Available from: https://ieeexplore.ieee.org/document/6486033

[20] Al-Shetwi et al. Active power control to mitigate frequency deviations in large-scale grid-connected PV system using grid-forming single-stage inverters. Energies. 2022;**15**:2035

[21] Polymeneas E, Tai H, Wagner A. Less Carbon Means More Flexibility: Recognizing the Rise of New Resources in the Electricity Mix. McKinsey Company. Available from: https://tinyurl.com/ywwb36e9; [Accessed: March 11, 2024]

[22] Rohmingtluanga C, Datta S, Sinha N. SCADA based intake monitoring for improving energy management plan: Case study. Energy Reports. 2023;**9**:402-410

[23] Abdolrasol MG, Hannan M. Optimal PI controller based PSO optimization for PV inverter using SPWM techniques. Energy Reports. 2021;**8**:1003-1011

[24] Chakraborty MR, Dawn S, Saha PK, Basu JB. A comparative review on energy storage systems and their application in deregulated systems. Batteries. 2022;**8**:124

[25] Tappeta VSR, Appasani B, Patnaik S. A review on emerging communication and computational technologies for increased use of plug-in electric vehicles. Energies. 2022;**15**:6580

[26] Kikusato H et al. Flywheel energy storage system based microgrid controller design and PHIL testing. Energy Reports. 2022;**8**:470-475

[27] Yadav K, Malik H. Case study of grid-connected photovoltaic power system installed at monthly optimum tilt angles for different climatic zones in India. IEEE Access. 2021;**9**:60077-60088. DOI: 10.1109/ACCESS.2021.3073136

[28] Pattnaik S, Kumar MR, Mishra SK, Gautam SP, Appasani B. DC bus voltage stabilization and SOC management using optimal tuning of controllers for Supercapacitor based PV hybrid energy storage system. Batteries. 2022;**8**:186

[29] Chauhan A, Upadhyay S, Khan MT. Performance investigation of a solar photovoltaic/diesel generator based hybrid system with cycle charging strategy using BBO algorithm. Sustainability. 2021;**13**:8048

[30] Dey PP, Das DC, Latif A. Active power management of virtual power plant under penetration of central receiver solar thermal-wind using butterfly optimization technique. Sustainability. 2020;**12**:6979

[31] Abdolrasol GM et al. Energy management scheduling for microgrids in the virtual power plant system using artificial neural networks. Energies. 2021;**14**:6507

[32] Pati A, Adhikary N, Mishra SK, Appasani B. Fuzzy logic based energy management for grid connected hybrid PV system. Energy Report. 2022;**8**:751-758

[33] Ulutas A, Altas IH, Onen A. Neuro-fuzzy-based model predictive energy management for grid connected microgrids. Electronics. 2020;**9**:900

[34] Hussain SMS et al. IEC 61850 based energy management system using plug-in electric vehicles and distributed generators during emergencies. International Journal of Electrical Power & Energy Systems. 2020;**119**:105873

[35] Jones JS. Futureproofing the Utility of the Future with IEC 61850. Smart Energy International. Available from: https://tinyurl.com/3cjyabms; 2023 [Accessed: March 11, 2024]

[36] Hussain SMS. Optimal energy routing in microgrids with IEC 61850 based energy routers. IEEE Transactions on Industrial Electronics. 2020;**67**(6):5161-5169

[37] Ustun TS, Hussain SMS, Orihara D, Iioka D. IEC 61850 modeling of an AGC dispatching scheme for mitigation of short-term power flow variations. Energy Reports. 2022;**8**(1):381-391

[38] Hussain SMS et al. A method for achieving confidentiality and integrity in IEC 61850 GOOSE messages. IEEE Transactions on Power Delivery. 2020;**35**(5):2565-2567

[39] Ustun TS, Farooq SM, Hussain SMS. Implementing secure routable GOOSE and SV messages based on IEC 61850-90-5. IEEE Access. 2020;**8**:26162-26171

[40] Farooq SM et al. Certificate based security mechanisms in vehicular ad-hoc networks based on IEC 61850 and IEEE WAVE standards. Electronics. 2019;**8**:96

[41] Farooq SM et al. Certificate based authentication mechanism for PMU communication networks based on IEC 61850-90-5. Electronics. 2018;**7**:370

[42] Hussain SMS. Smart inverter communication model and impact of cybersecurity attack. In: 2020 IEEE International Conference on Power Electronics, Drives and Energy Systems (PEDES); Jaipur, India. 2020. pp. 1-5. Available from: https://ieeexplore.ieee.org/document/9379762

[43] Unsal DB et al. Enhancing cybersecurity in smart grids: False data injection and its mitigation. Energies. 2021;**14**:2657

[44] Ustun TS, Hussain SMS, Ulutas A, Onen A, Roomi MM, Mashima D. Machine learning-based intrusion detection for achieving cybersecurity in smart grids using IEC 61850 GOOSE messages. Symmetry. 2021;**13**:826

[45] Ustun TS, Hussain SMS, Yavuz L, Onen A. Artificial intelligence based intrusion detection system for IEC 61850 sampled values under symmetric and asymmetric faults. IEEE Access. 2021;**9**:56486-56495. DOI: 10.1109/ACCESS.2021.3071141

[46] Ustun TS, Hussain SMS. A review of cybersecurity issues in Smartgrid communication networks. In: 2019 International Conference on Power Electronics, Control and Automation (ICPECA); New Delhi, India. 2019. pp. 1-6. Available from: https://ieeexplore.ieee.org/abstract/document/8975629

[47] Ustun TS, Hussain SMS. An improved security scheme for IEC 61850 MMS messages in intelligent substation communication networks. Journal of Modern Power Systems and Clean Energy. 2020;**8**(3):591-595

[48] Ustun TS. Cybersecurity vulnerabilities of smart inverters and their impacts on power system operation. In: 2019 International Conference on Power Electronics, Control and Automation (ICPECA); New Delhi, India. 2019. pp. 1-4. Available from: https://ieeexplore.ieee.org/document/8975537

[49] Ustun TS, Hussain SMS. IEC 62351-4 security implementations for IEC 61850 MMS messages. IEEE Access. 2020;**8**:123979-123985

[50] Farooq SM, Hussain SMS, Iqbal A. Using ID-based authentication and key agreement mechanism for securing communication in advanced metering infrastructure. IEEE Access. 2020;**8**:210503-210512

[51] Latif A, Hussain SMS, Das DC. Optimization of two-stage IPD-(1+I) controllers for frequency regulation of sustainable energy based hybrid microgrid network. Electronics. 2021;**10**:919

[52] Farooq Z, Rahman A, Hussain SMS. Power generation control of renewable energy based hybrid deregulated power system. Energies. 2022;**15**:517

[53] Nayak SR, Khadanga RK, Panda S, Sahu PR, Padhy S. Participation of renewable energy sources in the frequency regulation issues of a five-area hybrid power system utilizing a sine cosine-adopted African vulture optimization algorithm. Energies. 2023;**16**:926

[54] Barik AK, Das DC, Latif A, Hussain SMS. Optimal voltage–frequency regulation in distributed sustainable energy-based hybrid microgrids with integrated resource planning. Energies. 2021;**14**:2735

[55] Hussain I, Das DC, Sinha N, Latif A. Performance assessment of an islanded hybrid power system with different storage combinations using an FPA-tuned two-degree-of-freedom (2DOF) controller. Energies. 2020;**13**:5610

[56] Sahu PR, Lenka RK, Khadanga RK, Hota PK, Panda S. Power system stability improvement of FACTS controller and PSS design: A time-delay approach. Sustainability. 2022;**14**:14649

[57] Kikusato H et al. Aggregate modeling of distribution system with multiple smart inverters. In: 2019 International Conference on Smart Energy Systems and Technologies (SEST); Porto, Portugal. 2019. Available from: https://ieeexplore.ieee.org/document/8849085

[58] Ustun TS, Hashimoto J, Otani K. Impact of smart inverters on feeder hosting capacity of distribution networks. IEEE Access. 2019;**7**:163526-163536

[59] Driscoll W. In California and Hawaii, the Benefits of Smart Inverters Are Just Beginning. USA: PV Magazine; 2021

[60] Kikusato H et al. Developing power hardware-in-the-loop based testing environment for volt-var and frequency-watt functions of 500 kW photovoltaic smart inverter. IEEE Access. 2020;**8**:224135-224144

[61] Australian Energy Market Commission. Mandatory Primary Frequency Response, ERC0274. Brisbane QLD; 2020

[62] Hashimoto J et al. Development of df/dt function in inverters for synthetic inertia. Energy Reports. 2023;**9**(1):363-371

[63] Patil GS, Mulla A, Dawn S. Profit maximization with imbalance cost

improvement by solar PV-battery hybrid system in deregulated power market. Energies. 2022;**15**:5290

[64] Basu JB, Dawn S, Saha PK, Chakraborty MR. Economic enhancement of wind–thermal–hydro system considering imbalance cost in deregulated power market. Sustainability. 2022;**14**:15604

[65] Latif A, Paul M, Das DC. Price based demand response for optimal frequency stabilization in ORC solar thermal based isolated hybrid microgrid under Salp swarm technique. Electronics. 2020;**9**:2209

[66] Basu JB, Dawn S, Saha PK, Chakraborty MR. A comparative study on system profit maximization of a renewable combined deregulated power system. Electronics. 2022;**11**:2857

[67] Singh NK, Koley C, Gope S. An economic risk analysis in wind and pumped hydro energy storage integrated power system using meta-heuristic algorithm. Sustainability. 2021;**13**:13542

[68] Hussain SMS, Farooq SM. Implementation of Blockchain technology for energy trading with smart meters. In: 2019 Innovations in Power and Advanced Computing Technologies (i-PACT); Vellore, India. 2019. pp. 1-5. Available from: https://ieeexplore.ieee.org/document/8960243

[69] IEA. Unlocking the Potential of Distributed Energy Resources. Paris: IEA; 2022. Available from: https://www.iea.org/reports/unlocking-the-potential-of-distributed-energy-resources . Licence: CC BY 4.0 [Accessed: March 11, 2024]

Chapter 2

Post-Combustion Carbon dioxide (CO_2) Capture Technologies for Light Vehicle Internal Combustion Engines: A Technical Review of Present Trends, Challenges, and Prospects towards Reducing Global Carbon Footprint

Promise C. Anyanwu

Abstract

In the past decade, global attention has focused on global warming and the resultant climate change occasioned by the massive release of anthropogenic greenhouse emissions, mainly carbon dioxide (CO_2) into the atmosphere. To achieve net-zero targets by 2050, it is necessary to decarbonize high carbon-polluting sectors such as road transportation. According to several reports, internal combustion (IC) engine-driven road vehicles are among the biggest CO_2 emitters globally; hence, the deployment of novel carbon capture and storage (CCS) technologies in the sector is considered as a significant measure for scaling down carbon footprint. This review discusses various carbon capture technologies, with emphasis on post-combustion capture techniques. It further highlights on solid adsorbent materials that hold promising potentials for adsorption-based CO_2 capture and scholarly works that focus on their synthesis, modification, and use. The present trends in the development of solid adsorbent-based mobile post-combustion CO_2 capture devices for IC engine-driven road vehicles are discussed. The review also elucidates the current barriers to the economic-scale implementation of the current technologies. Future prospects and suggestions for advancements are presented, with particular emphasis on the synthesis of low cost, nano-sized materials. The chapter concludes with key findings and the overall impacts of implementing CCS technologies.

Keywords: carbon dioxide (CO_2), climate change, global warming, internal combustion engine, road transport, adsorption, post-combustion capture, nanomaterials

1. Introduction

Modern day internal combustion (IC) engine created by Nicolaus Otto in 1876 is unarguably one of the mankind's greatest inventions in history. As implied by the name, mixed fuel/air combustion takes place within a combustion zone where motive power is equally developed and extracted. Over the years since its development, the IC engine has been the primary prime mover for road vehicles, marine boats, ships, some aircrafts, and many stationary applications. The engine is typically powered by fossil-based fuels such as gasoline, diesel, natural gas, and renewable biofuels, such as bioethanol and biodiesel.

For over a century since its invention and subsequent improvements, the internal combustion engine has been a major driving force of industrialization and has played a significant role in societal development. However, with rising global energy costs and environmental concerns, there are campaigns from various quarters for a global ban and total phase out of IC engines. This is mainly due to the harmful emissions from their operation, which are detrimental to health and the environment. In recent times, global warming and the resultant climate change effects have gained serious attention and triggered actions from various segments. Many consider the adoption of electric vehicles (EVs) and other battery-powered renewable technologies as solutions for sustainable eco-friendly transportation in the future. Some experts also view the adoption of low-carbon fuels for aviation and marine operation and adoption of policies that promote more of public transportation as vital tools that will help achieve global climate goals.

The focus of this chapter is on road transport vehicles, which are widely considered as principal sources of hydrocarbon emissions, particularly in urban centers. Road transportation alone accounts for about 12% of greenhouse gas emissions globally [1]. Depending on the fossil fuel consumed, typical motor vehicle tailpipe emission may include particulate matter (PM), volatile organic compounds, nitrogen oxides (NO_x), carbon monoxide (CO), sulfur dioxide (SO_2), and carbon dioxide (CO_2). Several generations of catalytic conversion and other emission-reducing systems have been developed and deployed on vehicles over the past decades. These systems are designed to limit pollutants, such as CO, NO_x, and PM, in conformance to various European emission standards. However, despite these advancements in the exhaust system, CO_2 remains an end product of fossil fuel combustion. As a dominant greenhouse gas, which contributes significantly to global warming and climate change, the sequestration, safe storage, and recycling of CO_2 has been a key subject of both scientific and socio-economic attention globally. According to Tiseo [1], transportation generally accounts for 20% of global CO_2 emissions and is the second-largest carbon-polluting sector worldwide behind coal power generation. Over three quarters of this amount is from road transportation alone, representing about 15% of global CO_2 emissions. Out of this figure, passenger vehicles including cars, busses, and vans are the biggest emitters with a combined contribution of about 54%. The contribution of trucks conveying freight stands at about 25%. These numbers are quite significant and largely underpin the importance of reducing road transport-associated CO_2 emission. This will play a huge role as the world takes decisive actions in addressing climate change (**Figure 1**).

According to the United Nations Paris Agreement signed in 2016, there is a global target of achieving net-zero carbon emission by 2050 [2]. Based on current industry trends, expected technological advancements, and forecasts, CO_2 emission from the

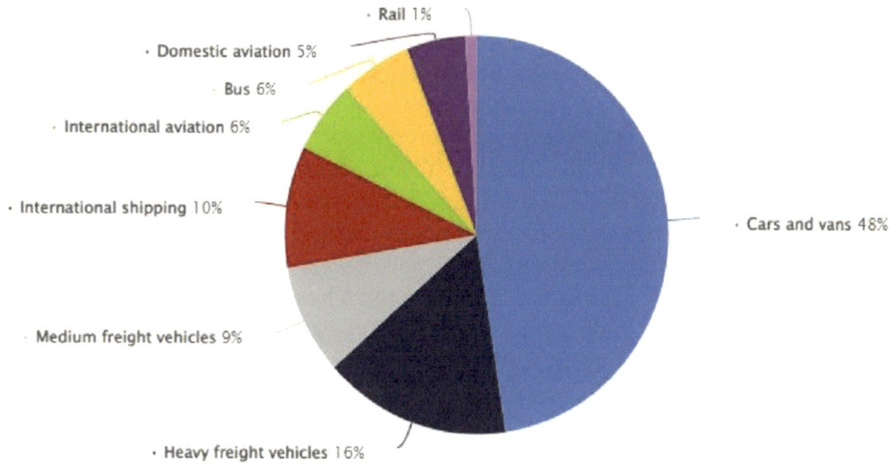

Figure 1.
Distribution of CO_2 emissions from global transportation in 2022 (source: [1]).

road transport is projected to peak by 2025 [1]. With increase in the global population, comes the rise in vehicular traffic [3]. The International Energy Agency [4] in its recent energy technology perspectives report projected a 100% increase in transportation, with a 60% rise in private vehicle ownership by the year 2070. This is in consideration of the present rate of global population growth. It also forecasted a 200% increase in passenger and freight airlifting. A close look at these figures shows that an increase in transport emissions is envisaged, indicating that the sector would still be a major contributor to energy-related emissions by the projected year.

Although developed nations have higher percentage of private-owned vehicles, air pollution and traffic gridlocks are worse in the developing and underdeveloped world [5]. Also, while developed nations are advancing in the adoption of green technologies and other policies to combat climate change, the rest of the world is lagging behind in implementing the same. Some of the countries struggle to formulate strategies for cost-effective mass transportation systems that will pose less detrimental effects on the environment. Given present socio-economic scenarios and realities in these parts of the world, fossil-fueled vehicles are envisaged to remain dominant in the foreseeable future; hence, the task of fully decarbonizing road transportation would be a tall order to accomplish by 2050. Interestingly, the substantial reduction of energy-related CO_2 emission using various carbon capture and storage (CCS) technologies has been highlighted in various reports. However, given the hugely significant yearly contribution of road transportation to global CO_2 emission, onboard mobile carbon capture represents a vital component for achieving net-zero targets. Alexander et al. [6] suggest that the integration of CCS technology in the transport sector could be an excellent way to keep the existing internal combustion engines in the market under the future net-zero CO_2 emission scenarios with a potential phase out of fossil fuels. Unlike power plants in which CCS systems are installed in the stationary mode, the adoption of the onboard mobile technology for road vehicles must consider its adaptation to IC engine operating dynamics, such as variations in mass flow and species concentrations in the flue gas, due to varying engine speed and load. In this direction, several studies [7–9] have emphasized the need to combine CCS technology with cleaner (less carbon-intensive) fuels that produce low CO_2 and particulate matter emissions to reduce their impact on the operation of the system.

Presently, there are efforts toward developing efficient CO_2 capturing devices for IC engine-driven road vehicles. This technical review, therefore, aims at ascertaining the achievements of various researchers in this direction. First, a brief overview of IC engine CO_2 emissions and the greenhouse effect is presented. Then, brief discussions on CO_2 capture technologies, with emphasis on post-combustion capture techniques follow. A review of solid adsorbents for adsorption-based CO_2 capture is then presented, highlighting previous and ongoing research works that focus on their synthesis and modification for improved performance. Thereafter, a review of various patents and research studies on mobile CCS technology is presented. The challenges of the existing systems, prospects, and suggestions for advanced work are further discussed. Generally, this review chapter aims to contribute to the growing research efforts toward the advancement, cost-effective, and full-scale implementation of mobile CO_2 capture technology on IC engine-driven road vehicles as an approach to combat climate change.

2. Internal combustion engine emissions

Flue gases or combustion products from IC engines are pollutants that distort the natural composition of the atmosphere. Data provided by the World Health Organization [10] indicate that the majority of the world's population inhale air containing these harmful substances, which exceed stipulated safe levels, particularly in developing and underdeveloped nations. Some of the substances are toxic and hence portend great danger to human health. They are known to cause severe respiratory problems, skin disorders, visual impairments, intestinal diseases, infant deformities, and other ailments, giving significant rise to mortality rates across the globe [11]. For instance, a 2020 estimate indicates that about 3.2 million deaths per year are attributable to air pollution, over 7% of which are children under the age of five [10]. Generally, the health effects associated depend on the nature of the substance, concentration, extent of exposure, and a person's immunity level. Aside posing serious threats to human health, some of the emitted substances are heat-trapping in nature, hence contribute adversely to global warming, which consequently gives rise to climate change. Typical internal combustion engine tailpipe emissions may include particulate matter (PM) and greenhouse gases such as carbon dioxide (CO_2), carbon monoxide (CO), nitrogen oxides (NO_x), sulfur dioxide (SO_2), volatile organic compounds (VOCs), polycyclic aromatic hydrocarbons, and hazardous heavy metals, such as lead [12]. Although the other gases may be considered harmful with their respective global warming potentials, emphasis mainly lie on carbon dioxide (CO_2), being the most dominant greenhouse gas emission in the atmosphere (**Figure 2**).

2.1 Greenhouse effect of carbon dioxide (CO_2) emission

Carbon dioxide is a primary combustion end product along with water vapor and a very important component of the carbon cycle. However, it remains the largest contributor to global warming and driver of climate change with a net positive radiative forcing estimated at about $3.7 Wm^{-2}$ [14]. Radiative forcing is defined the difference in the solar radiation absorbed by the earth and the energy radiated back to

Figure 2.
Internal combustion engine tailpipe emissions (source: [13]).

space. It is a measure of the influence of a given climatic change factor on the amount of downward radiant energy that reaches the earth's surface, as measured in watts per square meter. As of May 2022, the average global concentration of CO_2 in the atmosphere was 421 ppm (parts per million) [15]. This is a massive increase compared to its value at the start of the industrial revolution era, estimated at about 280 ppm [16]. Several anthropogenic (human) activities are responsible for this rise, mainly fossil fuel combustion. The problem has been exacerbated by the indiscriminate burning of green vegetation (bushes and forests) and falling of trees, which are designated by nature to be CO_2 sinks. In the past decades, post-combustion CO_2 capture and storage (CCS) has been a major issue for global attention and research. It is considered as a critical technology for reaching net-zero emissions.

3. CO_2 capture technologies

Depending on the application, CO_2 capture can be achieved by three main approaches, namely, pre-combustion, oxy-fuel combustion, and post-combustion [17]. In the pre-combustion capture process, the hydrocarbon fuel undergoes catalytic steam reformation (gasification) to produce synthetic gas composing of carbon monoxide (CO) and hydrogen (H_2). CO undergoes a further reaction with steam (referred to as shift reaction) to produce CO_2 and more hydrogen gas. The CO_2 is then trapped in a sorbent, while the hydrogen may be used as green fuel in a combustion turbine or a fuel cell application. Oxy-fuel combustion capture involves burning with pure oxygen rather than air, so as to maximize CO_2 yield and facilitate easier capture. Post-combustion capture involves the use of sorbents to separate CO_2 from the exhaust gas produced from the combustion process. It is easily adaptable and offers the advantage of being retrofitted to existing combustion systems in power plants and industrial facilities [18].

3.1 Post-combustion CO_2 capture

The main techniques for post-combustion CO_2 capture include amine absorption, membrane separation, cryogenic separation, carbonate looping, and solid sorbent adsorption [19, 20]. The liquid amine absorption is the most popular and widely adopted technique, especially in power plants due to the affinity of CO_2 to amine solvents [20–22]. However, the technique lends itself to several limitations such as high-energy requirements for regeneration, high capital intensity, chemical instability, toxicity, and corrosive tendency of the process equipment [23, 24]. These factors greatly limit its application for mobile CO_2 capture, although it has been proposed as a viable option for marine application [25–27].

Membrane separation involves isolating CO_2 from flue gas streams using high CO_2-selective membranes. The technique was originally deployed as an alternative to amine scrubbing in natural gas processing, but presently has extended applicability [28]. This is due to advantages such as low capital cost, less space requirements, and energy consumption [29, 30]. The operating process is simple and environmentally friendly as opposed to amine-absorption process. During the process, the membrane functions as a filter or molecular sieve, allowing only CO_2 molecules to permeate, while other molecules are restricted. The process utilizes characteristics such as chemical affinity/reactivity and kinetic diameter of the molecules [28]. Cellulose acetate and polymers (polycarbonates, polyamides, etc.) are common membrane materials that have been deployed for post-combustion CO_2 capture [31]. However, there are recent advancements regarding the application of non-polymeric materials such as activated carbon, zeolites, silica, and metal-organic frameworks (MOFs) for membrane-based post-combustion CO_2 capture [28]. The major challenge, with adopting this technology for mobile applications, such as IC engine, is the low CO_2 volumetric fraction in the flue gas that is approximately 15%, which results in insufficient pressure required to drive optimum CO_2 permeation [31]. Other limiting factors include chemical degradation, vapor accumulation, and a potential reduction in operational stability due to plasticization [32].

Cryogenic separation utilizes low temperatures to isolate and purify CO_2 from flue gas. The technology involves a series of thermal processes (compression and expansion) that condenses and distills liquid CO_2 or precipitates frozen CO_2 from the flue gas stream. This is achieved by first cooling and drying the flue gas stream within a range of very low temperatures that allows for easy separation of CO_2 from other constituents of the stream. It is an effective CO_2 capture technique that offers high CO_2 recovery rates and purity levels [33]. The overall capture efficiency, however, depends on temperature and pressure at the end of the expansion process [34]. The technology is straightforward and requires no chemicals or reagents. It produces readily conveyable CO_2 and is suitable for industrial-scale applications with large volumes of CO_2. However, the drawbacks that limit its use in mobile capture application include its relatively higher capital and operational costs and possibility of process blockage by other frozen constituents such as water, NO_x, and SO_x [35].

Carbonate looping is a post-combustion CO_2 capture technique that basically involves a reversible chemical reaction between metal oxides and the CO_2 contained in the flue gas to produce metal carbonates. Calcium carbonate looping is the most developed and commonly applied form of this technology. The process encompasses two main stages: carbonation and calcination [36]. The carbonation stage involves CO_2 adsorption or chemisorption on the calcium oxide (CaO) adsorbent to produce calcium carbonate ($CaCO_3$). The reaction is exothermic and occurs rapidly. The reverse stage calcination (CO_2 desorption) is slower and largely endothermic. It involves the thermal decomposition of the calcium carbonate to yield solid calcium oxide and CO_2. The near-pure stream of CO_2 is then extracted and purified for storage or use, while the calcium oxide adsorbent is regenerated for the subsequent cycle. It is noteworthy that while the process can be theoretically repeated infinitely, the calcium oxide sorbent degrades as it is recycled [36]. Due to this, it is necessary to purge out some of the used sorbent from the system and replace it with fresh sorbent. The size of the purge stream compared with the amount of sorbent going round the cycle affects the process considerably [37]. The calcium carbonate looping technology offers some advantages compared to conventional amine-absorption systems such as reduced energy penalty and the use of a cheap, widely available, and environmentally friendly adsorbent [38]. The drawbacks of the technology, which also limit its use for mobile application, include high capital cost of installation and rapid sorbent deactivation [39]. The latter is a significant decline in the sorbent's adsorption capacity and reactivity over a number of carbonation-calcination cycles, requiring a constant flow of fresh adsorbent in order to maintain the desired CO_2 capture efficiency [37, 40]. This obviously increases the operational and maintenance costs of the technology.

Generally, the above technologies are mainly suited to industrial applications, such as power plants. Thus, for small to medium scale applications, such as road vehicles, increased focus is on the deployment of the adsorption technique. This has given rise to the development of solid adsorbent materials, which possess excellent carbon capture capabilities. In the case of road vehicles, a promising approach is to directly trap the emission from source using retrofitted tailpipe emission capture devices developed with suitable adsorbent materials.

3.2 Solid adsorbent materials

Solid adsorbents are excellent and attractive candidates for point-source post-combustion carbon capture due to their less energy requirements and eco-friendliness [41–44]. They offer other advantages such as low cost, high availability, good textural quality, hydrophobicity, long-term stability, and easy regeneration [45–48]. They are basically classified as high-, medium- or low-temperature materials based on their thermal capabilities and operating temperatures [17, 49, 50]. They can also be classified as physical or chemical adsorbents depending on their nature. Physical solid adsorbents are typically porous materials such as zeolites, metal-organic frameworks (MOFs), covalent organic frameworks (COFs), porous polymer networks (PPN), silica, alumina amine-doped materials, and carbon-based materials, such as graphene, carbon nanotubes, carbon fibers, and activated carbon from bio-resources [21, 23, 51]. They are basically low-temperature materials (below 200°C), and adsorption of CO_2 on their pores at ambient temperature is on the basis of van der Walls and polar interactions [52, 53]. The capture performance of the physical adsorbent is influenced by the textural characteristics of the walls of the pores [54–56]. In some instances, to resist the effect of gas selectivity and enhance CO_2 capture, the physical adsorbent may be doped with an amine solvent, which has excellent chemical affinity with the gas [53, 55]. On the other hand, chemical adsorbents consist of oxides of metals such as calcium oxide (CaO), aluminum oxide (Al_2O_3), and titanium oxide (TiO_2) [34]. Others include hydrotalcites, metal silicates, and zirconates [21]. The adsorption temperatures of chemical sorbents are typically higher than their physical counterparts. Also, in contrast to the adsorption process with physical sorbents, CO_2 capture is on the basis of chemical bonding (chemisorption) between the molecules of the gas and the chemical sorbent [54, 55, 57]. The process is reversible, and the CO_2 adsorption/desorption cycle is temperature and pressure dependent [18, 58]. During the process, either a pressure swing or temperature swing separation technique is adopted in isolating CO_2 from other constituents of the flue gas [59, 60]. The high enthalpy of automobile exhaust gas favors the adoption of the temperature swing adsorption (TSA) technique. However, concerns have been raised about the limitations of the process, such as low loading capacity (mass of CO_2 captured per mass of material) of the sorbents. This requires that a sizeable amount of material must be provided to achieve more quantities of captured CO_2. Also, some of the chemical sorbent materials may be hydrophilic in nature. This results in a reduction in adsorption capacity, especially in instances where the exhaust gas has high-water content [61].

3.3 Functional characteristics of solid adsorbent materials

To effectively trap or isolate CO_2 from the exhaust gas, solid adsorbents must possess a range of important functional characteristics. These will ensure that they meet the required technical and economic standards for cost-effective post-combustion CO_2 capture, under a variety of working conditions. The most important characteristics, which constitute vital selection criteria, are discussed as follows [20, 62, 63]:

3.3.1 Adsorption capacity

This refers to the maximum amount of the gas (adsorbate) than can be adsorbed by the sorbent material at equilibrium state (adsorption rate equals desorption rate at the solid-gas interface) under given conditions of temperature and pressure. It is an important characteristic and selection criterion, which is influenced by the specific surface area, pore size, and surface reactivity of the adsorbent.

3.3.2 Gas selectivity

This is the most vital attribute of a physical adsorbent and refers to the degree to which the required adsorbate (CO_2) can permeate and be separated from other constituents of the flue gas stream. It is a measure of the effectiveness of an adsorbent in trapping CO_2, especially in cases involving multicomponent gaseous mixtures, such as flue gas. It is influenced by the pore size, pore geometry, pore distribution, and the material's affinity for CO_2. It determines the quality/purity of CO_2 adsorbed, which gives idea about the cost-effectiveness of the capture installation.

3.3.3 Reaction kinetics

This refers to the rate at which the reaction takes place. It determines the speed of the adsorption/desorption cycle and is influenced by the surface characteristics of the sorbent, molecular size of CO_2, concentration of CO_2 in the flue gas stream, and the kind of bonding that takes place. Ideally, for an excellent adsorbent material, reaction kinetics should be significantly fast to ensure less residence time required to reach interfacial equilibrium and promote fast regeneration.

3.3.4 Pore characteristics

This refers to the shape and size of the pores, which directly influence the effectiveness of the adsorption process. The relative size of CO_2 molecules is about 2 nm; hence, effective primary adsorption is achievable within the nanopores and micropores. The mesopores and macropores can serve as mechanism for CO_2 transport within the adsorbent.

3.3.5 Chemical and thermal stability

In most instances, solid adsorbents have to endure harsh chemical conditions during operation. Typically, the exhaust gas stream contains harmful compounds such as carbon monoxide, nitrogen oxides (NO_x), and volatile organic compounds (VOCs) that may degrade the strength or integrity of the material. They may also be subjected to extreme temperatures and pressures. Thus, it becomes very necessary that they possess robust stability and exceptional resilience under these harsh conditions of operation to ensure longevity, maintain high adsorption capacity, and continuous regeneration and reuse.

3.3.6 Sound mechanical properties

The adsorbent should possess excellent tensile and compressive strengths, hardness, and toughness in order to resist the effects of abrasion and crushing during its operation. These properties directly influence the efficiency of the process and the overall cost of the installation.

3.3.7 Ease of surface functionalization/enhancement

To improve the adsorption capacity and/or CO_2 affinity of the sorbent material, it may require surface functionalization or enhancement. This typically involves doping it with an amine-based solvent which contains a vital functional group. The chemical nature of the surface comes handy in this respect. The outer surface may also be functionalized to be hydrophilic (polar), while the internal structure remains hydrophobic (non-polar). This is a particularly important feature in the case of contact with flue gas streams with high-water content.

3.3.8 Regeneration capacity

Following the adsorption cycle in which CO_2 completely fills the pore volume and surface of the adsorbent, desorption (removal) of CO_2 molecules takes place in order to regenerate the adsorbent and restore it back for another cycle of operation.

3.3.9 Reliability and durability

The operational stability during cycles and the longevity of solid adsorbent materials are key selection criteria, which influence the frequency of material change and an indication of the cost-effectiveness of the installation.

3.3.10 Low cost

This is an important techno-economic consideration in developing adsorbent materials for CO_2 capture. The material should be easily accessible and capable of being produced with minimal cost. This will impact on its commercial scalability.

3.3.11 Environmental friendliness

In an era of environmental consciousness, it is important that used adsorbent materials do not pose toxic effects or cause ecological harm when disposed.

4. Nano-sized materials for mobile post-combustion CO_2 capture

The functional characteristics elucidated above are the general requirements for effective post-combustion CO_2 capture for an adsorbent. However, reducing the size of material particles to the nanoscale provides unique physical and chemical attributes that favors post-combustion CO_2 capture. Generally, nano-sized particles allow for unique interactions that are not practically achievable at larger scales. The attributes include high surface area to volume ratio for increased chemical reaction and enhanced adsorption capacity, high surface energy, spatial confinement with minimal structural imperfections, high thermal stability, high mechanical strength, etc. In addition, they are highly energy-efficient, economical, and eco-friendly [64]. As an emerging area, research works that focus on their use for CCS applications are relatively few in literature; hence, the gains derivable from these favorable attributes have not been fully explored. However, it is believed that the purposeful synthesis of carbon nanotubes, nanoporous silica, graphene, and zeolite nanoparticles among others for mobile post-combustion capture devices could offer significant improvement in terms of capture performance and process efficiency. There is also a growing interest in the use of nanostructured separation membranes for mobile CO_2 capture. They are considered to be highly energy-efficient (utilize only about 8% of energy generated), and their low cost can significantly cut down overall cost of installation [65]. Available forms include polymer nanostructured membrane, carbon molecular sieve membranes, and ceramic nanostructured membranes.

4.1 Zeolite nanoparticles

Presently, zeolites are among the most economically available materials for post-combustion carbon capture. They possess a tetrahedral crystalline structure, which basically consists of aluminum and silica. This framework allows for a variety of pore structures [51]; hence, they can be considered microporous, mesoporous, or nanoporous. Several kinds of zeolite materials with modifications, such as HZSM-5 and M-ZSM-5, have been deployed as sorbents in direct air CO_2 capture (DAC) technology [66]. Generally, zeolites are popular capture materials for several factors, including high selectivity for CO_2 uptake, regularized pore structure with interconnected channels, large surface area, modifiable textural properties, and molecular sieve effect [67]. However, despite these positives, they have inherent challenges, which limit their advanced application. A major constraint is the high energy required for quick desorption and regeneration owing to strong adhesion of CO_2 on their pore surface. Also, their hydrophobic (water-sensitive) nature is a limiting factor for high adsorption efficiency and quick regeneration, especially in applications involving flue gas with high-water content. However, despite these shortfalls, they remain viable candidates for the development of low-cost tailpipe CO_2 capture devices due to their wide availability. With advancements in CO_2 capture technology, the synthesis of nano-sized zeolite particles could offer some improvement in terms of performance (**Figures 3** and **4**).

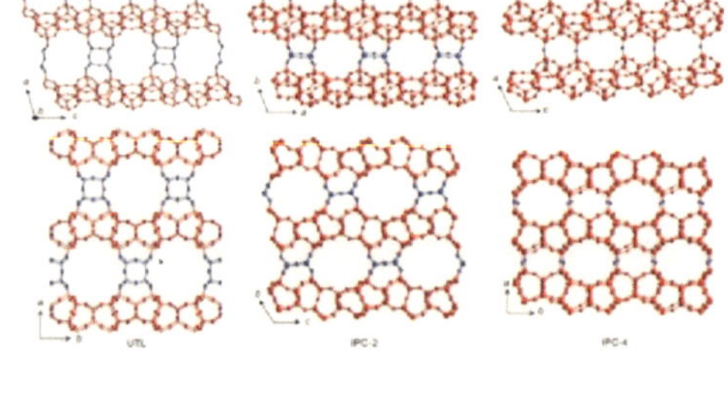

Figure 3.
Different zeolite pore structures (source: [68]).

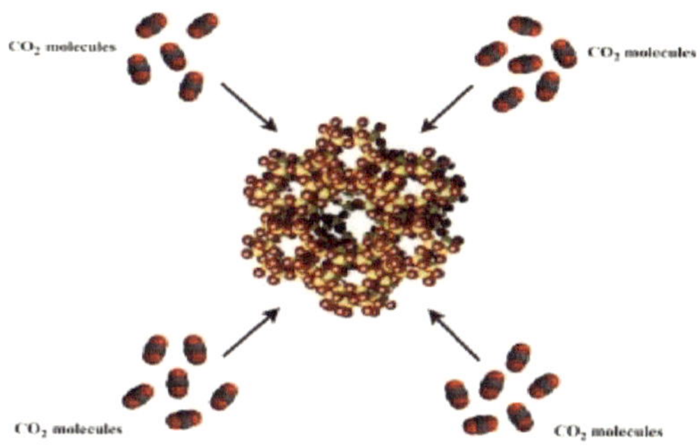

Figure 4.
Mechanism of CO_2 adsorption on zeolite particles (source: [20]).

Zeolites are among the most researched materials for adsorption-based CO_2 capture owing to their wide availability. Research efforts have focused on addressing their performance limitations through modification or composite synthesis. Ullah et al. [69] studied CO_2 adsorption on six different kinds of modified zeolite adsorbents comprising of three ZSM-5 (CBV-2314, CBV-3024E, and CBV-5524G with Si-Al ratio of 23, 30, and 50, respectively), zeolite-Y (CBV-300), zeolite-β (CP814E), and mordenite (CBV-21A) at high pressure of 200 bar and at different temperatures of 35, 50, and 65°C. Experimental result showed that zeolite-β had the highest CO_2 uptake of 72 mmol/g at 35°C, although it exhibited relatively lower selectivity for the adsorbate. However, ZSM-5 (CBV-2314) recorded the lowest CO_2 capture of 19 mmol/g under similar test conditions. The researchers attributed the high CO_2 adsorption recorded for zeolite-β over other sorbents to the large Si-Al ratio and its pore structure and volume. Similarly, the performance of amine-functionalized zeolite-β was investigated by Madden et al. [70]. The adsorbent was doped with amino-propyltriethoxysilane to reduce its surface area to about 23 m^2/g from 454 m^2/g. The test results showed very significant improvement in CO_2 adsorption capacity and enhanced material stability after repeated cycles, even under wet conditions. The stability was attributed to the presence the silicon group in the doping agent, which attaches to the pore surface of the parent material. This also lowered the CO_2 desorption temperature (60°C), indicating that the adsorption/desorption process is dependent on the position and pore structure of the functional group. Another study by Panda et al. [71] compared the CO_2 uptake efficiency of zeolite HZ4A-13 under dry and wet conditions. Test results showed that the material achieved a 34.67% increase in CO_2 uptake compared to the dry state at 0.15 bar. This is indicative of its enhanced performance in limiting the adverse effects of water in the flue gas at low pressure. The material also exhibited superior CO_2 uptake efficiency when compared to zeolite 5A and zeolite 13X; hence, it was proven to be a viable option for wet flue gas CO_2 capture.

4.2 Graphene and graphite nanoparticles

Graphene is a basic carbonaceous material with a two-dimensional honeycomb-like lattice structure. It possesses good characteristics for CO_2 capture, which includes large surface area, high adsorption capacity, electrical conductivity, mechanical stability, and ease of surface modification. However, graphene has less functional groups and its strong intermolecular interaction limits its compatibility and ease of dispersal [72]. Thus, it is mostly deployed in alternate forms, such as graphene oxide, which possess better properties [20]. Graphite is another low cost and commercially available option for post-combustion CO_2 capture. Its molecular structure consists of graphene layers connected by weak van der Waals forces. CO_2 adsorption between the layers is made possible by moderating their spatial distances. However, due to its weak affinity for CO_2 and small surface area for sufficient adsorption, it is best suited as base material in the synthesis of composites, surface-activated, or doped with an amine-based solvent to improve its CO_2 uptake (**Figure 5**) [20].

Several studies on modified forms of graphite and graphite oxide for enhanced CO_2 capture are reported in literature. Mishra et al. [73] showed that graphite nanoplatelets have good CO_2 capture potential at elevated working pressures. Hong et al. [74] functionalized graphite oxide with 3-aminopropyl-triethoxysilane to enhance its CO_2 affinity and adsorption capacity. The material also exhibited high thermal stability during the operational cycles. Similarly, Dominguez et al. [75] employed sodium dodecyl sulfate as doping agent to enhance CO_2 adsorption on a graphite surface. Zhao et al. [76] studied the performance of various urea-doped graphite materials for CO_2 capture. The material with the highest nitrogen fraction showcased superior performance (CO_2 uptake of 4.23 mmol/g at 25°C) under dry conditions. Similar studies on modified graphene have also been reported. Xia et al. [77] modified the structure of porous graphene for enhanced CO_2 capture through chemical activation using potassium hydroxide (KOH), CO_2, and graphite oxide. Tests conducted at 0°C and 1 bar yielded a peak adsorption value of 4.06 mmol/g. Another popular modification of graphene-based adsorbents is attaching organic functional groups to the graphene surface. Amine functionalization has also been employed to enhance the functional characteristics of graphene for enhanced CO_2 capture. In a study by Hosseini et al. [78], graphene oxide samples modified with three different amine solvents, including triethylenetetramine (TETA), meta-phenylenediamine (MPD), and pentaethylenehexamine (PEHA), were tested at ambient conditions (25°C and 1 bar). Results showed significant improvements in CO_2 uptake and selectivity over N_2 for the three samples. In summary, it may be noted that surface functionalization plays a significant role in boosting the effectiveness of both graphite and graphene for CO_2 capture.

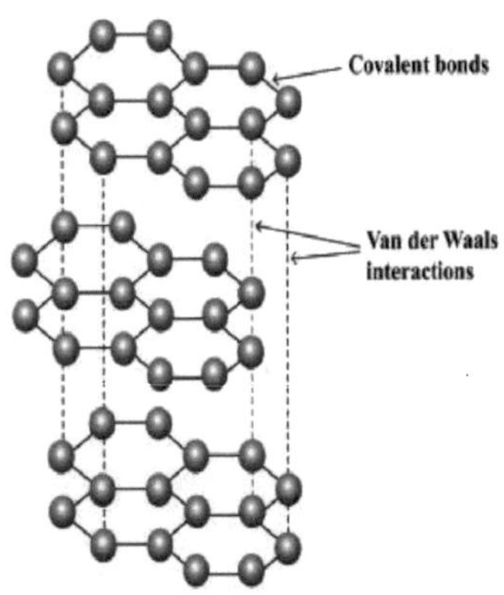

Figure 5.
Structure of graphite (source: [20]).

4.3 Carbon nanotubes (CNTs)

Carbon nanotubes are considered as suitable post-combustion capture materials because they possess appropriate surface properties, hollow tube-shaped structures, good tensile strength, low weight, thermal and chemical stability, and good pore-size distribution [79]. They also possess easily modifiable surfaces for enhanced CO_2 uptake. CNTs can be very long and have cross sectional diameters up to tens of nanometers. Their internal structure consists of several graphene layers rolled up in form of a hollow tube. This geometry allows the distribution of the adsorbate and residence time to be easily controlled. CO_2 adsorption rate in carbon nanotubes is determined by gas pressure, temperature, their diameter in the total material volume, and the distance between adjacent graphene layers [80]. They are classified based on the number of graphene layers that make up their wall, a feature which determines their mechanical and thermal potentials. These include single-walled carbon nanotubes (SWCNTs), double-walled carbon nanotubes (DWCNTs), and multi-walled carbon nanotubes (MWCNTs), each of which have proven effectiveness for specific gas applications (**Figure 6**) [20].

Several studies have shown amine-functionalized CNTs to be very viable, non-toxic, and energy-efficient options for CO_2 capture from flue gas streams. Keller et al. [81] activated MWCNTs with polyethyleneimine solvent (PEI). Lower proportions of the solvent enhanced the surface area and porosity, while CO_2 adsorption increased with increase in the amount of the solvent, reaching a peak value of 2.11 mmol/g (at 20% by weight of solvent). It, however, decreased with proportions of the solvent above 20%. Lee et al. [82], in a similar study, attributed the enhanced sorption efficiency to the attachment of amino groups, which creates oxygen affinity sites on the free carbon surface. Alternatively, Rahimi et al. [83] modified MWCNTs using 1,3-diaminopropane, which improved CO_2 adsorption from 1.10 to 2.11 mmol/g.

4.4 Silica nanoparticles

Silica is a hard crystalline solid material basically consisting of silicon and oxygen, which are the most abundant elements on earth. Its potential for CO_2 capture is based on its well-defined structure. However, its amorphous form does not effectively support CO_2 adsorption. It lacks definite and favorable structure for pore control, surface enhancement, and effective binding of the gas molecules [20]. The unique attributes of porous silica materials have expanded their applications as solid adsorbents for CO_2

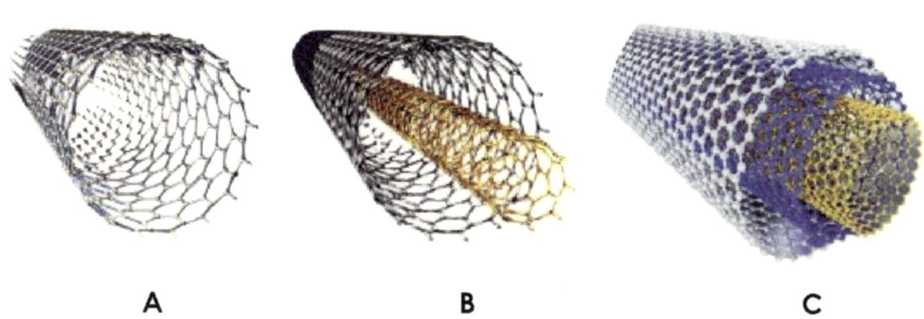

Figure 6.
(A) SWCNTs, (B) DWCNTs, and (C) MWCNTs. (Source: [20]).

capture in various applications. Their advantages include high adsorption capacity under high pressure (about 45 bar), ease of surface enhancement or modification, expanded surface area and pore volumes, adjustable pore size, low manufacturing cost, superb regeneration stability with low energy requirements, good gas selectivity, and good thermal and mechanical stability [84]. Their textural properties, such as surface area, pore volume, pore size, and distribution, can be easily modified for enhanced CO_2 uptake. However, a major drawback with their application, which is a subject of research, is their limited hydrothermal stability, which deals with the hydrolysis of their bonds. Silica-based materials possess the ability to accommodate nanoparticles of different metals within their pore structure, an attribute that favors the development of composites. They can also combine with other solid adsorbents, such as CNTs and metal-organic frameworks (MOFs) (**Figure 7**) [20].

Due to its viable potential as support material for synthesis and development of new adsorbents for post-combustion CO_2 capture, significant research is progressing in the direction. A comparative study by Kumar et al. [86] showed the performance of different amine-impregnated mesoporous silica adsorbents. The samples were prepared using three classes of nanoparticles having different pore configurations and doped with five different amine solvents, including ethylene diamine (EDA), diethylenetriamine (DETA), triethylenetetramine (TETA), tertaethylenepentamine (TEPA), and polyethylenimine (PEI). Tests were conducted under wet and dry conditions, and CO_2 adsorption capacities were recorded. An optimal amine loading of 50–60% by weight was documented for all the samples. Maximum adsorption capacities of 2.17 mmol/g and 2.10 mmol/g were recorded for sample SBA-15 containing 50–60% by weight of TETA and TEPA, respectively, under dry condition. Similarly, the lowest CO_2 adsorption capacities, 0.12 mmol/g and 0.50 mmol/g, were recorded for mesoporous silica doped with EDA and DETA, respectively. At wet conditions, peak value of 2.16 mmol/g was recorded for SBA-15 containing 50% by weight of polyethylenimine, while the least values were recorded for samples MCM-41 and MCM-48 due to poorly distributed amines within their pores. In a related study, Zhang et al. [87] investigated the capture performance of amine-functionalized mesoporous silica. CO_2 uptake recorded at different temperatures exhibited an increasing trend within the first 30 min of adsorption. A maximum CO_2 uptake of 2.3 mmol/g was obtained at 30°C. However, the value decreased significantly by 80% as the temperature rose from 30–90°C, indicating the effect of temperature on CO_2 adsorption on the sorbent.

4.5 Alumina nanoparticles

Alumina, also referred to as aluminum oxide (Al_2O_3), is another widely available and low-cost option that can be developed for CO_2 capture. It is presently a popular support material in a number of CO_2 capture applications and is available in several crystallographic forms. As a solid adsorbent, alumina offers the benefits of cost-effectiveness, large reactive surface area, and excellent porosity with large pore volumes,

Figure 7.
Powdered silica (source: [85]).

sound chemical, mechanical, and thermal stability [88]. It is a viable candidate for post-combustion capture devices; hence, several research works focus on improving its properties in order to enhance its CO_2 uptake, lower residence time and quicker adsorption-desorption cycle, and give high CO_2 selectivity over other gases contained in the flue gas (**Figure 8**).

Several research works on the synthesis alumina as base material for post-combustion CO_2 capture are reported in literature. Bararpour et al. [91] synthesized alumina-aerogel with potassium carbonate (K_2CO_3), which yielded a high stability and capacity material. The synthesis provided an improvement on other alumina-based materials with a significantly higher reactive surface area of 2019 m^2/g. Also, a significant CO_2 uptake of 7.2 mmol/g was recorded at 56.1°C. In another study [92], the researchers examined the effect of textural properties on CO_2 adsorption on mesoporous alumina synthesized with K_2CO_3, using a surfactant-functionalized technique. The material developed with surfactant to alumina ratio of 0.5, exhibited the maximum CO_2 uptake of 5.7 mmol/g. It maintained thermal stability during 15 adsorption/desorption cycles under test. The enhanced CO_2 adsorption capacity was linked to the presence of multiple active sites for chemical reaction.

4.6 Activated carbons (ACs)

Activated carbons are very effective solid adsorbents and among the cheapest and most widely available materials for post-combustion CO_2 capture from exhaust gas streams. This is because they are developed from huge amounts of waste biomass and industrial wastes, which possess low inorganic content, but high carbon composition. Aside being cost-effective, they offer functional benefits such as low desorption temperature requirement, good textural qualities for enhanced CO_2 uptake, fast regeneration, minimal energy consumption, quick adsorption-desorption cycle, high thermal and chemical stability, good tensile strength, and good thermal conductivity [93]. Activated carbons can also be synthesized from natural sources, such as coal and lignite [94]. Different raw material compositions create structural differences in the structure of activated carbons, such as porosity and surface area. This indicates that CO_2 capture on ACs can easily be modified. They mainly possess microporous and mesoporous structures with different pore shapes [95], although nanoporous ACs are being developed for enhanced CO_2 adsorption. An interesting feature is the dual surface polarity of activated carbons. The outer surfaces are hydrophilic (polar), while the internal structures are hydrophobic (non-polar). This characteristic provides

Figure 8.
Alumina granules and nano-powder (sources: [89, 90]).

versatility in the sequestration of various greenhouse gas emissions. During their production, the waste raw materials (feedstock) first undergo either thermal decomposition to produce biochar or hydrothermal conversion to produce hydrochar [20]. The solid products achieved typically exhibit poor porosity and a low specific surface area, which poses constraints for CO_2 capture [96]. Thus, to improve these properties, activation is performed. There are basically two activation methods involved in their synthesis: physical and chemical activation. Physical activation involves thermochemical conversion processes, such as pyrolysis, where the char products are partially gasified in the presence of oxidizing agents at an elevated temperature range (800–1000°C). Chemical activation, however, involves the reaction with chemical reagents and then carrying out heat treatment under neutral conditions. The process is achieved at a less temperature range (450–900°C) (**Figure 9**) [98].

A number of studies have been conducted on AC adsorbents from biomass for post-combustion CO_2 capture application. Parshetti et al. [99] developed microporous AC adsorbents from lignocellulose material obtained from empty palm nut. The preparation process involved pulverization and hydrothermal carbonization of the raw material in a reactor for 20 minutes at different temperatures (150, 250, and 350°C). The hydrochar materials produced were chemically activated with potassium hydroxide (KOH). The products were sampled for testing according to their carbonization and activation temperatures. The test results showed that adsorbent carbonized at 250°C and activated at 800°C (designated H250–800) possessed the highest specific surface area of 2511 m^2/g and total pore volume of 1.06 cm^3/g. It also exhibited the highest CO_2 adsorption capacities: 5.18, 3.71, and 2.34 mmol/g at 0, 25, and 50°C, respectively. Also, selectivity for CO_2 over N_2 was 11.2 at 25°C. The CO_2 uptake efficiency was maintained over 10 adsorption/desorption cycles. Some researchers have also investigated commercial-grade ACs as adsorbents for CO_2 capture. Sahoo et al. [100] compared seven commercially developed Acs, including Norit RB 4, Norit RB 4C, AC CARB 6 × 12 55, AC CARB 6 × 16 55, AC CARB 8 × 16 60, AC CARB 6 × 12 60, and Modisorb PL4 for post-combustion CO_2 capture. The test samples were highly microporous with specific surface areas ranging from 923 m^2/g to 1254 m^2/g, while total pore volume varied from 0.38 m^3/g to 0.58 m^3/g. The CO_2 adsorption capacities of the samples recorded at 1 bar and 0°C varied from 3.02 mmol/g to 4.53 mmol/g. The maximum value was recorded for AC CARB 6 × 12 55, while the least value was recorded for Norit RB 4. Also, the adsorption capacities of the samples varied from 2.15 mmol/g to 3.15 mmol/g at ambient conditions (25°C, 1 bar).

4.7 Nanoporous metal-organic frameworks (MOFs)

Metal-organic frameworks (MOFs) are a growing class of porous materials that have received interest in recent times. Their structures are highly modifiable, and this allows for the construction of functional derivatives with unique and desired

Figure 9.
Activated carbon particles (source: [97]).

properties suitable for applications in various areas such as energy, carbon sensing and capture, biotechnology, and catalysis. They are developed by self-assembly of metal ions and organic linkers and possess unique and favorable characteristics for CO_2 capture. These include a large surface area for adsorption, tuneable pore sizes, ease of surface enhancement, and a high selectivity for CO_2 [101]. Other advantages include low-pressure drop, rapid mass transfer kinetics, and very high storage capacity [49]. Despite these merits, MOFs have some shortfalls, which limit their adoption in certain applications. They are not very stable and can exhibit low uptake efficiency under low-pressure conditions, (ideal pressure range for CO_2 capture from flue gas), as well as when exposed to a mixture of gases under dynamic conditions [51]. Current research efforts are, therefore, focused on improving the adsorption capacity, selectivity toward CO_2, and stability of MOFs. There is a need for further understanding of their self-assembly and transformation mechanisms. **Figure 10** illustrates several nanoporous structures of MOFs as developed by various researchers.

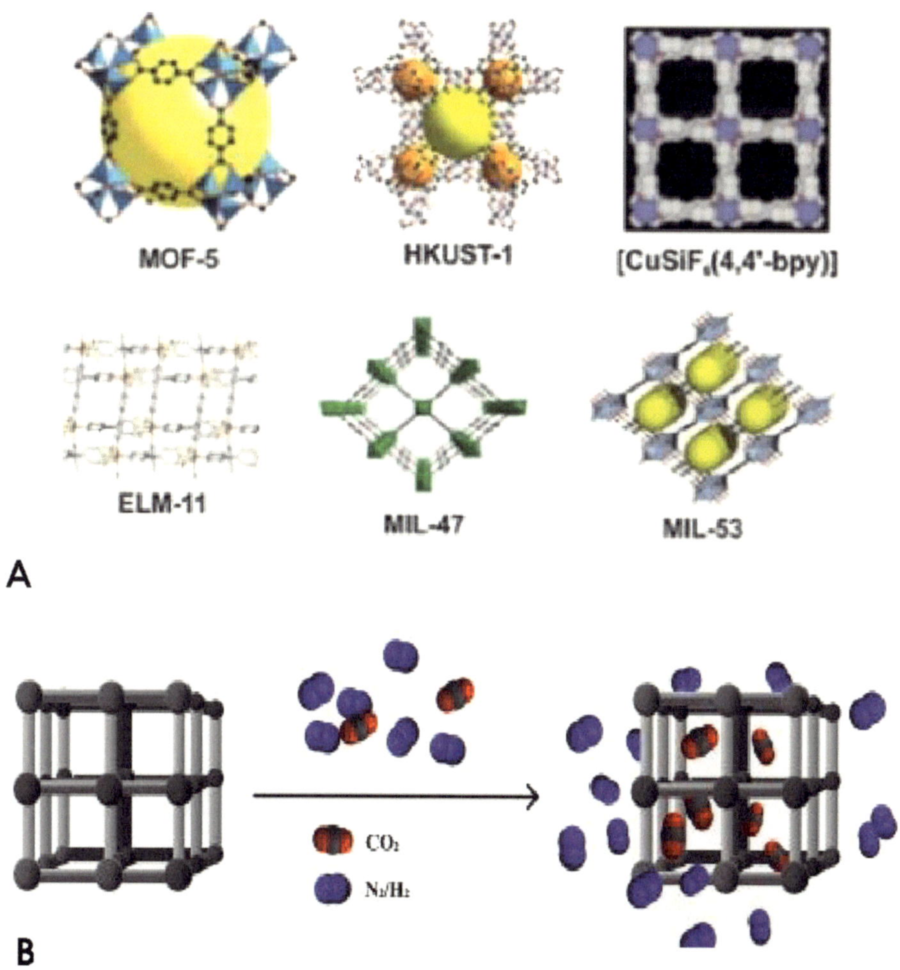

Figure 10.
A. Nanoporous MOF structures (source: [102]). B: Adsorption of CO_2 within nanoporous MOF structures (source: [20]).

Most of the research works detail on the influence of operational conditions and flue gas composition on CO_2 adsorption. This research mainly concentrates on the influence of moisture, as well as pollutants. Some other works focus on tuning pore shape and size, modifying metal sites, developing new structures, optimizing regeneration and recycling processes, pore surface functionalization, and synthesis of MOF composites. Wen et al. [103] developed a MOF-based solid adsorbent with suitable pore size. The material exhibited superior CO_2 adsorption capacity; 3.56 mmol/g at 0.15 bar and 23°C. It also exhibited high selectivity CO_2 over N_2 from flue gas stream. The values recorded were significant improvements on those recorded for other MOF structures reported. Qian et al. [104] improved the pore structure of carbon monolith (HCM) by combining the functional merits of MOF-based material ($Cu_3(BTC)_2$) in a bid to maximize high-volume CO_2 adsorption. The composite produced exhibited superior CO_2 adsorption capacity, high selectivity for CO_2 over N_2, and excellent regenerative ability.

4.8 Metal oxide nanoparticles

Metal oxides are potential chemical solid adsorbents for post-combustion CO_2 capture. The popular kinds include calcium oxide (CaO), magnesium oxide (MgO), titanium dioxide (TiO_2), and iron oxides (F_xO_x). They ideally react with the captured CO_2 to form metal carbonates. Adopting a particular kind of metal oxide for flue gas application will depend on its ability to form solid carbonates that possess thermodynamic stability during the process [105]. The initial capture phase occurs rapidly and involves the carbonation (CO_2 adsorption) of the adsorbent, which is an exothermic reaction. The later phase is slower and involves a decarbonation (desorption) process, which removes the trapped CO_2 and regenerates the adsorbent for subsequent adsorption. The process is largely endothermic (**Figure 11**).

Generally, metal oxides are considered as suitable options due to their favorable properties. Titanium dioxide is widely adopted for flue gas applications due to its strong affinity for CO_2, high adsorption capacity, and high gas selectivity [107]. Also, its large surface area and minimal pore sizes result in faster uptake kinetics, hence a shorter residence time for the adsorption process [108]. This feature is due to its highly regular pore channels, which significantly improve the diffusion rate, ensuring the rapid uptake of gas molecules during the adsorption process [109]. Calcium and iron oxides are recognized and widely used for their adsorption capacities (high CO_2 affinity), low cost, wide availability, excellent thermodynamic properties, durability, and ease of synthesis. In addition, calcium oxides are easily recyclable and less toxic, hence environmentally friendly and can be easily regenerated [110].

Figure 11.
Various samples of metal oxide nanoparticles (source: [106]).

Magnesium oxide has the advantage of consuming very low energy. However, it has inherent drawbacks, which may limit its applicability. It has a constrained surface area, which restricts CO_2 adsorption capacity, particularly at very elevated temperatures [111]. This also affects its recyclability and hence, makes it cost-ineffective for several applications. However, as CO_2 capture technologies evolve, there are renewed interests on the suitability of magnesium oxide for CO_2 adsorption, particularly at nanoscale, giving its relatively lower operational power requirement. Ruhaimi et al. [112] recently presented a comprehensive analysis on the deployment of MgO-based adsorbents developed with different methods of modification and synthesis.

Several other researchers have reported on the synthesis of metal oxides for CO_2 capture. Mendoza et al. [113] developed a novel technology for capturing CO_2 using different iron oxides (Fe_3O_4-Fe and Fe_2O_3-Fe) from steel processing. The work was based on CO_2 pressure and milling process parameters and involved examination of the kinetics of mechanically activated chemical reactions between the iron oxides and CO_2. They reported that carbonation of the oxides is achievable by applying the gas pressure (10–30 bar) at ambient temperature. At similar conditions of operation, the Fe_2O_3-Fe setup exhibited higher CO_2 adsorption than that of Fe_3O_4-Fe. Nawar et al. [114] explored the use of eggshells in modifying the structure CaO with the use of organic acids. The study introduced eggshell as a cost-effective material for CO_2 capture.

4.9 Porous organic polymer (POP) nanoparticles

Porous organic polymers (POPs) are novel classes of porous materials, composed of low-mass elements interconnected by covalent bonds. They have proven efficacy across diverse applications, including gas storage and separation [115]. Although several classes of POPs have been derived through research, the common classes associated with CO_2 capture include conjugated microporous polymers (CMPs), covalent organic frameworks (COFs), and hyper-crosslinked polymers (HCPs) [20]. They possess positive attributes, including low density, expanded surface areas, tuneable pore sizes with multiple channels, and abundant active sites available for chemical reactions. In addition, their surfaces can be easily functionalized with suitable organic groups. They also possess exceptional physicochemical stability, a wide array of design options, and synthetic approaches [116]. These factors make POPs strong candidates for point-source CO_2 capture (**Figure 12**).

In the case of post-combustion flue gas separation, advancements on POPs focus on enhancing their adsorption capacity at high temperatures and low pressures (increased number of active adsorption points for CO_2 binding), material stability in the presence of moisture and at high temperatures, cost-effectiveness, and economic scalability. Recently, Wang et al. [117] adopted thiophene as a base material for the development of new CMPs. The produced adsorbent exhibited excellent physicochemical stability and superior CO_2 adsorption capacity14.73 mmol/g at 50 bar and 44.85°C. Similarly, Dong et al. [118] employed ionic liquid for post-synthetic modification of a COF-based adsorbent. The improved material demonstrated superior CO_2 adsorption capacity of 3.74 mmol/g (1 bar, 0°C) and high physiochemical and thermal stability. Penchah et al. [119] studied the effects of process parameters (cross-linker ratio, synthesis time, and catalyst type) on CO_2 uptake of carbazole-based HCP, synthesized through a Friedel-crafts reaction. At a temperature of 25°C and a pressure of 5 bar, CO_2 adsorption reached a peak value of 4.35 mmol/g. In other studies, Fayemiwo et al. [120] deployed the bulk copolymerization technique

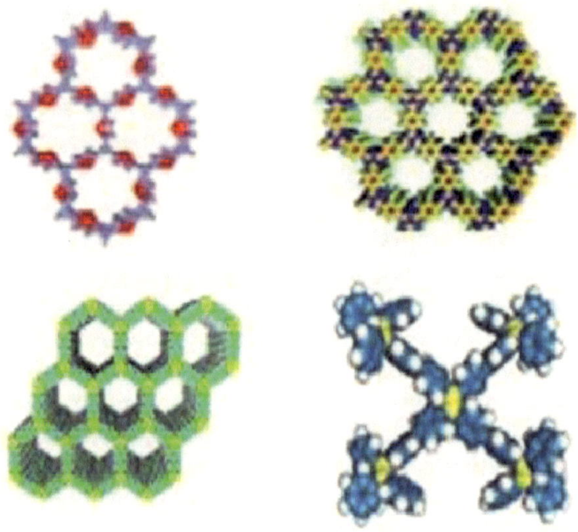

Figure 12.
Various POPs lattice structures (source: [20]).

in the development of nitrogen-rich, hyper-crosslinked polymers (HCP-MAAMs). The amine-functionalized products exhibited high selectivity for CO_2 over N_2 at very low gas partial pressure. The peak CO_2 adsorption capacity recorded at 0°C was 1.56 mmol/g.

5. Advances in IC engine CO_2 capture

Post-combustion CO_2 capture technology is yet to attain commercial viability; however, various researchers have developed patents and performed studies to evaluate their performances. Researchers at Stanford University [121] developed a post-combustion CO_2 capture device using a menaline-based porous network adsorbent. Their choice of menaline (a component of Formica) as base material was based on its low cost, easy accessibility, ease of manufacture, and energy efficiency with desirable CO_2 capture properties. It was modified with diethylenetriamine (DETA) to enhance CO_2 adsorption and desorption. DETA is a colorless, hygroscopic byproduct derived from the production of ethylene diamine from ethylene dichloride [122]. The material was further functionalized with cyanuric acid for enhanced porosity and regenerative capacity. The device captures CO_2 at about 40°C and releases it at about 80°C, offering significant energy savings. The capture efficiency of the device was compared to that of similar device developed with a novel metal-organic framework (MOF) adsorbent material, and it proved to be more efficient in trapping CO_2 from flue gas.

The studies [123–131] evaluated various solid adsorbent materials, mounted on exhaust tailpipes of spark ignition (SI) and compression ignition (CI) engines. The SI engines were fueled with gasoline, while the CI engines ran on conventional diesel and biodiesel, respectively. The solid adsorbents evaluated include zeolite, activated carbon, activated alumina, calcite, and monoethanolamine (MEA)-doped zeolite. Experimental results compared show that the devices with zeolite and MEA-doped zeolite as adsorbent materials yielded better performance by achieving up to 60% CO_2 capture rate. The results achieved were attributed to the high adsorption capacity of the zeolite sorbent, as well as CO_2's high affinity to amine solvents. Similarly, another solid sorbent capture device developed by Saudi Aramco [132] for their patented CCS technology achieved a 10% CO_2 capture rate when installed and tested on a Ford F-250 pick-up truck. However, physical confinement and mass/volume constraints prompted them to switch to a solvent-based system (potassium carbonate solution) for the subsequent phase of the project. This enabled them to shrink the entire system to about one-eighth of its previous size and confine the system under the chassis of a passenger vehicle. The later installation achieved an improved 30% CO_2 uptake efficiency under driving conditions.

Study conducted by Sharma et al. [59] showed the performance of their patented temperature swing adsorption-based CO_2 capture system for an internal combustion engine. The sorbent examined was a porous polymer network (PPN) material. The system was simulated in an organic Rankine cycle and achieved a capture rate of 90%. The result indicated that the energy required for CO_2 capture can be wholly dependent on the recovered waste heat from the exhaust gases. The study however did not consider the effect of load and speed variations, which could influence the capture efficiency of the system. Ahmed et al. [133] evaluated a newly developed system, which combines both absorption and adsorption techniques. Sodium oxide (NaO) was deployed as adsorbent, and the test was conducted on synthetic exhaust gas. The result showed a 100% CO_2 capture rate for duration of 70 minutes of operation but gradually declined to zero after about 220 minutes. Similar study by Muthiya et al. [18] showed the performance of their novel capture device with zeolite-based adsorbent, designed for a single-cylinder diesel engine. The test result indicated a very significant CO_2 adsorption by the device with minimal back pressure.

Giuseppe et al. [60] made significant progress with CO_2 capture from heavy-duty freight transport. Two novel metal-organic frameworks with high CO_2 selectivity and high storage capacity were respectively adopted: KAUST-7 and Al-soc-MOF-1. The former possesses high stability and selectivity toward CO_2, hence was selected as the capture medium. The later was selected as the storage medium given its high gravimetric and volumetric CO_2 uptake between 10 to 50 bar. They aimed at achieving CO_2 uptake efficiency of about 50%, with up to 95% CO_2 purity. The capture material's thermodynamic and kinetic properties were measured and modeled, and process parameters were simulated and optimized in response to engine dynamics. Captured and stored mass and volume were also minimized, representing a significant method of trapping tailpipe CO_2 from heavy-duty trucks (**Figure 13**).

Firm, Remora [134] also developed a tailpipe-mounted CO_2 capture device for heavy-duty haulage trucks. A zeolite-based adsorbent material was adopted in manufacturing the device, which operates on the temperature swing adsorption (TSA) principle. The frame-mounted assembly filters and isolates tailpipe CO_2 with

Figure 13.
Scientists developing a novel CO_2 capture device (source: [134]).

Figure 14.
Frame-mounted CO_2 capture assembly by remora (source: [134]).

an uptake efficiency of about 80%. Another firm, Qaptis [135] embarked on an ambitious drive to develop a very efficient CO_2 capture device for heavy-duty transport. According to their reports, the developed device captures up to 90% of CO_2 emission and can be retrofitted directly into the internal combustion engine. It utilizes waste heat from the exhaust gas for its operation (**Figure 14**).

6. Challenges of present technologies

Although several solid adsorbent materials have been developed and already have wide range applicability, further research works are ongoing to address their operational limitations. This is aimed at optimizing their CO_2 uptake efficiencies for diverse applications, such as mobile capture. For road vehicles, already developed patents exist but are yet to be fully commercialized, mostly due to their functional limitations and cost implications. It is noteworthy that the dynamic nature of internal combustion engine operation complicates the CO_2 capture process. This is due to variations in operating conditions, thermodynamic properties of the exhaust gas, mass flow variations, etc. Moreso, there are issues with vehicular space limitations. The mass and volume requirements for capture and storage materials remain critical challenges for implementing the adsorption-based post-combustion capture on road vehicles. These are very crucial factors, particularly with the use of carbon-intensive fuels, such as gasoline and diesel, which produce exhaust gases containing large volumes of CO_2. Also, the added energy required for powering the supplementary hardware and possible back pressure effect on engine performance arising from the capture installation are potential concerns for full-scale implementation. These challenges have established gaps for further research, hence the need to develop more technically and commercially viable facilities for internal combustion engine post-combustion CO_2 capture.

7. Prospects for future work

Research efforts are currently focused on developing feasible, inexpensive, high-capacity adsorbent materials for post-combustion CO_2 capture devices. The growing field of nanotechnology is the one that provides a promising route. Nanomaterials will play crucial roles in the future due to their excellent functional characteristics required for effective CO_2 capture. As a research direction, the purposeful characterization and synthesis of low cost and widely accessible nanomaterials will be instrumental in advancing the present technologies. The evaluation of prospective materials typically provides required data on their thermophysical and fluid-dynamic properties to ensure they satisfy technical and economic criteria that favor efficient CO_2 adsorption. It is important to develop novel CO_2 capture devices that can operate under a variety of process conditions and have high commercial scalability. Further, synthesis will be based on information on the practical limitations of existing materials, as presented in literature. It is noteworthy that developing materials that are able to satisfy all necessary functional criteria may be quite unusual and challenging; hence, it is important to obtain an appropriate balance that yields optimal results. During development, it is crucial to study the mechanism of potential adsorbent materials, which precisely informs about their interactions with CO_2, as well as CO_2's storage within the structures of the adsorbent. A method of achieving this is by integrating thermodynamic principles with computational simulations, data on their structural characteristics, sorption equilibrium, reaction kinetics, mass spectroscopy, and modeling the basic adsorption equations. This provides necessary validation, eliminates guesswork, and offers valuable information required for their scalability. It can also help to limit production time and costs. Developing new synthesis methods or approaches could also help in lowering final production costs. It is also necessary during analysis to have realistic trials for new products. For instance, it is important to conduct CO_2 capture tests under real flue gas conditions, such as low pressure, high temperature, and variant moisture levels. An important research direction involves the development of highly functional composite materials based on a variety of adsorbents, which yield unique properties for CO_2 adsorption from flue gases.

Furthermore, a techno-economic assessment is required as a crucial step that provides information on the technical and economic viability of the new products, with particular emphasis on their commercial scalability. The techno-economic assessment typically involves comparison of the potentially viable adsorbent materials for industrial purposes, outlining relevant specific criteria (benchmarks), technological variants, scalability, and stability costs.

8. Conclusion

As the world gears up for a potential future phase out of fossil-fueled vehicles, it is suggested that the transition to electric vehicles and hydrogen technologies should progress along with other technological innovations, which could offset emissions. The outcome of this review shows that appreciable mileage has been attained in the development of novel technologies for transport-associated CO_2 capture. It shows that there are promising prospects with the implementation of adsorption-based capture on road vehicles. However, some of the results have been obtained at either computational or laboratory scale and therefore remain subject to practical or real-life constraints. Some other systems have not been fully explored to consider the effects of operational variations, such as speed and engine load. These, therefore, open up avenue for extended research, with a view to optimize CO_2 capture rates using novel, modified, or improved materials, enhance the speed of the adsorption/desorption cycle, and surmount some of the present practical constraints. As new technologies for post-combustion CO_2 capture emerge, the kind of adsorbent to be adopted is a major factor for consideration. Specifically, the next generation adsorbent materials must exhibit a range of desired properties in order to be considered economically viable and environmentally sustainable for mobile capture application. This will also impact on the operating costs and energy requirement.

In summary, CO_2 capture or sequestration is among the technologies expected to address global energy challenges. Thus, the full-scale implementation of the mobile post-combustion CO_2 capture technology is of enormous benefit. It is important to note that the captured CO_2 can be channeled to industries for production of polymers and other chemicals. This will create high-value job opportunities, especially with the rapid commercialization of the products. The resultant effect will be a significant growth of the economy. More importantly, the technology is hugely significant toward scaling down global CO_2 emission. The gas is globally attested as the biggest climatic change factor. Although huge capital investments are required for advanced research and upscaling of the mobile capture technology, the economic gains are clearly huge, and it is crucial to the attainment of global climate goals.

Author details

Promise C. Anyanwu
Engineering Faculty, Mechanical Engineering Department, Ahmadu Bello University, Kaduna, Nigeria

*Address all correspondence to: prograce19@gmail.com

References

[1] Triceo N. Transportation Emissions. 2023. Available from: https://www.statista.com/topics/7476

[2] The United Nations (UN). The Paris Agreement. New York, United States: The United Nations (UN). Available from https://www.un.org/climate-change; 2023 [Accessed: September 30, 2023]

[3] Kharis H. Traffic, air pollution and health. In: Advances in Transportation and Health. Elsevier; 2020. pp. 59-104. DOI: 10.1016/B978-0-12-819136-1.00003-6

[4] International Energy Agency (IEA). Energy Technology Perspectives 2020. Paris: IEA; 2020

[5] Odogun A, Georgakis P. Transport pollution: A research of the Nigerian transport sector. International Journal of Innovative Technology and Exploring Engineering (IJITEE). 2019;**8**(11S):492-497. DOI: 10.35940/ijitee.K1083.09811S19. ISSN: 2278-3075

[6] Alexander G, Eva L. Review on carbon capture in ICE driven transport. Energies. 2021;**14**:6865. DOI: 10.3390/en14216865

[7] Kalghatgi G, Levinsky H, Colket M. Future transportation fuels. Progress in Energy and Combustion Science. 2018;**69**:103-105

[8] Osorio-Tejada J, Llera-Sastresa E, Scarpellini S. Liquefied natural gas: Could it be a reliable option for road freight transport in the EU? Renewable and Sustainable Energy Reviews. 2017;**71**:785-795

[9] Mikulcic H, Ridjan-Skov I, Dominkovic F, Wan-Alwi S, Manan Z, Tan R, et al. Flexible carbon capture and utilization technologies in future energy systems and the utilization pathways of captured CO_2. Renewable and Sustainable Energy Reviews. 2019;**114**:109338

[10] World Health Organization (WHO). Household Air Pollution. Geneva, Switzerland: World Health Organization (WHO). Available from https://www.who.int/news-room/fact-sheets/detail/household-air-pollution-and-health [Accessed: December 15, 2023]

[11] Donwa PA, Mgbame CO, Utomwen OA. Gas flaring in the oil and gas sector in Nigeria. International Journal of Commerce and Management Research. 2015;**1**(1):28-39

[12] Richard G, Sylvester C, Muhammad I. Air pollution in the Niger Delta region of Nigeria: Sources, health effects, and strategies for mitigation. Journal of Environmental Studies [JESJ]. 2023;**29**(1):1-15

[13] Paolo Bona/Shutterstock. 2022. Available from: https://www.shutterstock.com/g/Paolo+Bona (image source)

[14] Britannica. Radiative Forcing Potential of Carbon Dioxide Emissions. Chicago Illinois, United States: Encyclopedia Britannica; 2023. Available from: https://www.britannica.com/science/global-warming/Radiative-forcing [Accessed: October 5, 2023]

[15] Ramaswamy V, Boucher O, Haigh J, Hauglustaine D. Radiative Forcing. 2019. Available from: https://www.ipcc.ch/site/assets/uploads/2018/03/TAR-06.pdf 3

[16] Stanford doer school of sustainability. Concentration of CO_2 in

the Atmosphere. 2023. Available from: https://sustainability.stanford.edu/

[17] Orr FM Jr. CO_2 capture and storage: Are we ready? Energy & Environmental Science. 2009;**2**(5):449-458. DOI: 10.1039/B822107N

[18] Muthiya J, Kumar S, Amarnath V. Carbon capture and storage from automobile exhaust to reduce CO_2 emission. International Journal of Innovative Research in Science Engineering and Technology. 2014;**3**(2):33-37

[19] Pahija E, Golshan S, Blais B, Boffito D. Perspectives on the process intensification of CO_2 capture and utilization. Chemical Engineering and Processing Process Intensification. 2022;**176**:108958

[20] Bartosz D, Serafin J, Andersson K, Renata K. CO_2 capture materials: A review of current trends and future challenges. Materials Today Sustainability. London, UK: Elsevier Limited; 2023;**24**:100483. pp. 1-79. DOI: 10.1016/j.mtsust.2023.100483

[21] Christiano B, Pedro M, Leonel J, Leandro C. Advances in carbon capture and use (CCU) technologies: A comprehensive review and CO_2 mitigation potential analysis. Clean Technologies. 2022;**2022**(4):1193-1207. DOI: 10.3390/cleantechnol4040073

[22] Sharma S, Rangaiah G, Maréchal F. Integrated multi-objective differential evolution and its application to amine absorption process for natural gas sweetening. Differential Evolution in Chemical Engineering: Developments and Applications. 2017;**128**:-155. DOI: 10.1142/9789813207523_0005

[23] Serafin J, Mohammed O, Orlando F, Cruz J, Joanna S. Preparation of low-cost activated carbons from Amazonian nutshells for CO_2 storage. Biomass and Bioenergy. 2021;**144**:105925. DOI: 10.1016/j.biombioe.2020.105925

[24] Lin Q, Zhang X, Wang T, Zheng C, Gao X. Technical perspective of carbon capture, utilization, and storage. Engineering. 2022;**14**:27-32

[25] Luo X, Wang M. Study of solvent-based carbon capture for cargo ships through process modelling and simulation. Applied Energy. 2017;**195**:402-413

[26] Awoyomi A, Patchigolla K, Anthony EJ. CO_2/SO_2 emission reduction in CO_2 shipping infrastructure. International Journal of Greenhouse Gas Control. 2019;**88**:57-70

[27] Awoyomi A, Patchigolla K, Anthony EJ. Process and economic evaluation of an onboard capture system for LNG-fueled CO_2 carriers. Industrial and Engineering Chemistry Research. 2020;**59**:6951-6960

[28] Petros G, Efrosini P, Anastasios Z. Membrane-based technologies for post-combustion CO_2 capture from flue gases: Recent progress in commonly employed membrane materials. Membranes. 2023;**13**(12):898. DOI: 10.3390/membranes13120898

[29] Kim S, Scholes CA, Heath D, Kentish S. Gas-liquid membrane contactors for carbon dioxide separation: A review. Chemical Engineering Journal. 2021;**41**:128468

[30] Asghar U, Rafiq S, Anwar A, Iqbal T, Ahmed A, Jamil F, et al. Review on the progress in emission control strategies for the abatement of CO_2, SO_x and NO_x from fuel combustion. Journal of Environmental Chemical Engineering. Amsterdam, Netherlands:

Elsevier B.V Limited; 2021;**9**(5):106064. DOI: 10.1016/j.jece.2021.106064

[31] Guozhao J, Ming Z. Membrane separation Technology in Carbon Capture. In: Recent Advances in Carbon Capture and Storage. London, UK: IntechOpen; 2016. Available from: https://www.intechopen.com/chapters/53169. DOI: 10.5772/65723

[32] Scholes C. Challenges for CO_2 capture by membranes. In: Advances in Carbon Capture Methods, Technologies and Applications. London, UK: IntechOpen; 2020. pp. 357-377. DOI: 10.1016/B978-0-12-819657-1.00016-5

[33] Carolina F, David C, Chinonyelum U. Review of cryogenic carbon capture innovations and their potential applications. C. 2021;7(3):58. DOI: 10.3390/c7030058

[34] KAPSARC. CCE Technologies – Cryogenic Carbon Capture (Online resource). Riyadh, Saudi Arabia: King Abdullah Petroleum Studies and Research Centre (KAPSARC); 2024. Available from: https://www.cceguide.o rg/technologies/remove/manufacturing-industry/carbon-capture/cryogenic-carbon-capture [Accessed February 15, 2024]

[35] Nayef G. Cryogenic process: CO2 removal from natural gas. In: Advances in Carbon Capture. Elsevier Science; 2020. DOI: 10.1016/C2018-0-05339-6

[36] Blamey J, Anthony E, Wang J, Fennell P. The calcium looping cycle for large-scale CO_2 capture. Progress in Energy and Combustion Science. 2010;**36**(2):260-279. DOI: 10.1016/j.pecs.2009.10.001

[37] Rodriguez N, Alonso M, Grasa G, Abanades J. Heat requirements in a calciner of $CaCO_3$ integrated in a CO_2 capture system using CaO. Chemical Engineering Journal. 2008;**138**(1-3):148-154. DOI: 10.1016/j.cej.2007.06.005

[38] Dumitru C, Ioana I. Biomass CO_2-firing with carbon capture. Comprehensive Renewable Energy (Second Edition). 2022;**5**:330-347. DOI: 10.1016/B978-0-12-819727-1.00044-3

[39] Dong J, Tang Y, Nzihou A. Effect of steam addition during carbonation, calcination or hydration on re-activation of CaO sorbent for CO_2 capture. Journal of CO_2 Utilization. 2020;**3**:101167. DOI: 10.1016/j.jcou.2020.101167

[40] Filipe R, Sérgio G, Adrian B, Pinheiro C. Artificial neutral network modelling for the prediction of the deactivation of CaO-based adsorbents in the calcium looping process for CO_2 capture. Computer Aided Chemical Engineering. 2023;**52**:1259-1265. DOI: 10.1016/B978-0-443-15274-0.50201-8

[41] Rashidi NA et al. An overview of activated carbons utilization for the post-combustion carbon dioxide capture. Journal of CO_2 Utilization. 2016;**13**:1-16

[42] Prauchner MJ et al. Chemical versus physical activation of coconut shell: A comparative study. Microporous and Mesoporous Materials. 2012;**152**:163-171

[43] Chomiak K et al. Optimizing the properties of granular walnut-shell based KOH activated carbons for carbon dioxide adsorption. Journal of CO_2 Utilization. 2017;**21**:436-443

[44] Serafin J et al. Direct conversion of biomass to nanoporous activated biocarbons for high CO_2 adsorption and supercapacitor applications. Applied Surface Science. 2019;**497**:143722

[45] Sabouni R et al. Carbon dioxide adsorption in microwave-synthesized

metal organic framework CPM-5: Equilibrium and kinetics study. Microporous and Mesoporous Materials. 2013;**175**:85-91

[46] Hao W et al. Activated carbons prepared from hydrothermally carbonized waste biomass used as adsorbents for CO2. Applied Energy. 2013;**112**:526-532

[47] Pevida C et al. Silica-templated melamine–formaldehyde resin derived adsorbents for CO_2 capture. Carbon. 2008;**46**:1464-1474

[48] Akinola T, Bonilla P, Wang M. Experimental studies, molecular simulation and process modelling/simulation of adsorption-based post-combustion carbon capture for power plants: A state-of-the-art review. Applied Energy. 2022;**317**:119156

[49] Wang Q, Luo J, Zhong Z, Borgna A. CO_2 capture by solid adsorbents and their applications: Current status and new trends. Energy & Environmental Science. 2011;**4**:42-55. DOI: 10.1039/c0ee00064g

[50] Akeeb O, Wang L, Xie W, Davis R, Alkasrawi M, Toan S. Post-combustion CO_2 capture via a variety of temperature ranges and material adsorption process: A review. Journal of Environmental Management. 2022;**313**:115026. DOI: 10.1016/j.jenvman.2022.11502

[51] Olusola O, Helen M, Oluwatayo R, Chinemerem R, Christianah A, Peter O. (2023) adsorbent technologies and applications for carbon capture, and direct air capture in environmental perspective and sustainable climate action. Sustainable Chemistry for Climate Action. 2023;**3**:100029. DOI: 10.1016/j.scca.2023.100029

[52] Øi L, Procedia SK-E. Comparison of energy consumption for different CO_2 absorption configurations using different simulation tools. Energy Procedia. 2014;**63**:1186-1195

[53] Cabriga C, Clarete K, Zhang J, Pacia R, Ko Y, Castro J. Evaluation of biochar derived from the slow pyrolysis of rice straw as a potential adsorbent for carbon dioxide. Biomass Conversion and Biorefinery. 2021;**1**:3

[54] Gunawardene O, Gunathilake C, Vikrant K, Amaraweera S. Carbon dioxide capture through physical and chemical adsorption using porous carbon materials: A review. Atmosphere. 2022;**13**:397

[55] Gunathilake C, Manchanda A, Ghimire P, Kruk M, Jaroniec M. Amine-modified silica nanotubes and nanospheres: Synthesis and CO_2 sorption properties. Environmental Science: Nano journal. 2016;**3**:806-817

[56] Yao M, Wang L, Hu X, Hu G, Luo M, Fan M. Synthesis of nitrogen-doped carbon with three-dimensional mesostructures for CO_2 capture. Journal of Materials Science. 2015;**50**:1221-1227

[57] Vinodh R, Babu C, Abidov A, Palanichamy M, Jang H. Facile synthesis of amine modified silica/reduced graphene oxide composite sorbent for CO2 adsorption. Materials Letters. 2019;**247**:44-47

[58] Farmahini A, Krishnamurthy S, Friedrich D, Brandani S, Sarkisov L. Performance-based screening of porous materials for carbon capture. Chemical Reviews. 2021;**121**:10666-10741

[59] Sharma S, Francois M. Carbon dioxide capture from internal combustion engine exhaust using temperature swing adsorption. Frontiers in Energy Research. 2019;**7**:143. DOI: 10.3389/fenrg.2019.00143

[60] Giuseppe P, Abdulhadi A, Adrian R, Carlos A, Jorge G, Mohamed E, et al. Onboard capture and storage system using metal-organic frameworks for reduced carbon dioxide emissions from vehicles. Cell Reports Physical Science. 2023;**4**(7):10146. DOI: 10.1016/j.xcrp.2023.101467

[61] Zanco S, Joss L, Hefti M, Gazzani M, Mazzotti M. Addressing the criticalities for the deployment of adsorption-based CO_2 capture processes. Energy Procedia. London, UK: Elsevier; 2017;**114**:2497-2505. DOI: 10.1016/j.egypro.2017.03.1407

[62] Yang R. Gas Separation by Adsorption Processes. Vol. 1. London, UK: Imperial College Press; 1997. DOI: 10.1142/p037

[63] Rouquerol J, Rouquerol F, Llewellyn P, Maurin G, Sing KS. Adsorption by Powders and Porous Solids: Principles, Methodology and Applications. Academic Press; 2013. DOI: 10.1016/C2010-0-66232-8

[64] Abu-Dief M, Salaheldeen M. Opinion on nanomaterials for carbon capture and conversion. Materials Science and Engineering. 2020;**4**(6):149-150. DOI: 10.15406/mseij.2020.04.00142

[65] Cuffe L, MacElroy JMD, Tacke M, Kozachok M, Mooney DA. The development of nanoporous membranes for separation of carbon dioxide at high temperatures. Journal of Membrane Science. 2006;**272**(1/2):6-10

[66] Choi S, Drese J, Jones C. Adsorbent materials for carbon dioxide capture from large anthropogenic point sources. ChemSusChem. 2009;**2**:796-854. DOI: 10.1002/cssc.200900036

[67] Lai J, Ngu L. A review of CO_2 adsorbents performance for different carbon capture technology processes conditions. Greenhouse Gases: Science and Technology. 2021;**11**:1076-1117. DOI: 10.1002/ghg.2112

[68] Roth W, Nachtigall P, Morris R, et al. A family of zeolites with controlled pore size prepared using a top-down method. Nature Chemistry. 2013;**5**:628-633. DOI: 10.1038/nchem.1662 (image source)

[69] Ullah R, Saad M, Aparicio S, Atilhan M. Adsorption equilibrium studies of CO_2, CH_4 and N_2 on various modified zeolites at high pressures up to 200 bars. Microporous and Mesoporous Materials. 2018;**262**:49-58. DOI: 10.1016/j.micromeso.2017.11.022

[70] Madden D, Curtin T. Carbon dioxide capture with amino-functionalized zeolite-β: A temperature programmed desorption study under dry and humid conditions. Microporous and Mesoporous Materials. 2016;**228**:310-317. DOI: 10.1016/j.micromeso.2016.03.041

[71] Panda D, Kumar E, Singh S. Introducing microporosity in zeolite 4A bodies for rapid CO_2 capture. Journal of CO2 Utilization. 2020;**40**:101223. DOI: 10.1016/j.jcou.2020.101223

[72] Stoller M, Park S, Zhu Y, An J, Ruoff R. Graphene-based ultra-capacitors. Nano Letters. 2008;**8**(10):3498-e3502. DOI: 10.1021/nl802558y

[73] Mishra K, Ramaprabhu S. Study of CO_2 adsorption in low cost graphite nanoplatelets. IJCEA. 2010;**1**(3):266-e269. DOI: 10.7763/IJCEA.2010.V1.46

[74] Hong S, Kim H, Lee B. Adsorption of carbon dioxide on 3-aminopropyl-triethoxysilane modified graphite oxide. Energy & Fuels. 2013;**27**(6):3358-3363. DOI: 10.1021/ef400467w

[75] Dominguez H, Espinosa-Jimenez H. CO_2 adsorption on a modified graphite

surface with sodium dodecyl sulfate surfactants: A molecular dynamics study. Revista Mexicana de Fisica. 2019;**65**(1):20-24

[76] Zhao Y, Seredych M, Zhong Q, Bandosz T. Superior performance of copper based MOF and laminated graphite oxide composites as CO_2 adsorbents at room temperature. ACS Applied Materials & Interfaces. 2013;**5**(11):4951-4959. DOI: 10.1021/am404952z

[77] Xia K, Xiong R, Chen Y, Liu D, Tian Q, Gao Q, et al. Tuning the pore structure and surface chemistry of porous graphene for CO_2 capture and H_2 storage. Colloids and Surfaces A: Physicochemical and Engineering Aspects. 2021;**622**:126640. DOI: 10.1016/j.colsurfa.2021.126640

[78] Hosseini Y, Najafi M, Khalili S, Jahanshahi M, Peyravi M. Assembly of amine-functionalized graphene oxide for efficient and selective adsorption of CO2. Materials Chemistry and Physics. 2021;**270**:124788. DOI: 10.1016/j.matchemphys.2021.124788

[79] Smart S, Cassady A, Lu G, Martin D. The biocompatibility of carbon nanotubes. Carbon. Amsterdam, Netherlands: Elsevier B.V Limited; 2006;**44**(6):1034-1047. DOI: 10.1016/j.carbon.2005.10.011

[80] Belin T, Epron F. Characterization methods of carbon nanotubes: A review. Materials Science and Engineering B. 2005;**119**(2). DOI: 10.1016/j.mseb.2005.02.046

[81] Keller L, Ohs B, Lenhart J, Abduly L, Blanke P, Wessling M. High capacity polyethylenimine impregnated microtubes made of carbon nanotubes for CO_2 capture. Carbon. 2018;**126**:338-345. DOI: 10.1016/j.carbon.2017.10.023

[82] Lee M, Lee Y, Park J. Preparation and characterization of multi-walled carbon nanotubes impregnated with polyethyleneamine for carbon dioxide capture. International Journal of Hydrogen Energy. 2015;**40**(8):3415-3421. DOI: 10.1016/j.ijhydene.2014.12.104

[83] Rahimi K, Riahi S, Abbasi M, Fakhroueian Z. Modification of multi-walled carbon nanotubes by 1, 3-diaminopropane to increase CO_2 adsorption capacity. Journal of Environmental Management. 2019;**242**:81-89. DOI: 10.1016/j.jenvman.2019.04.036

[84] Belmabkhout Y, Serna-Guerrero R, Sayari A. Adsorption of CO_2 from dry gases on MCM-41 silica at ambient temperature and high pressure. 1: Pure CO_2 adsorption. Chemical Engineering Science. San Francisco, United States: Elsevier Inc.; 2009;**64**(17):3721-3728. DOI: 10.1016/j.ces.2009.03.017

[85] Asbestos council of Victoria. What Is Silica. 2023, Asbestos council of Victoria. Available from: https://gards.org/what-is-silica/ (image source)

[86] Kumar R, Bandyopadhyay M, Pandey M, Tsunoji N. Amine-impregnated nano-architectonics of mesoporous silica for capturing dry and humid 400 ppm carbon dioxide: A comparative study. Microporous and Mesoporous Materials. 2022;**338**:11956. DOI: 10.1016/j.micromeso.2022.111956

[87] Zhang B, Liu X, Wang H, Li K, Tatsumi T, Wang J. Accessible amino-rich ordered mesoporous silica for carbon dioxide capture. Materials Letters. 2022;**321**:132422. DOI: 10.1016/j.matlet.2022.132422

[88] Seki T, Onaka M. Mesoporous alumina: Synthesis, characterization, and catalysis. Advanced Nanomaterials.

2009;**2009**:481-521. DOI: 10.1002/9783527628940.ch15

[89] Smart Elements. Activated Alumina (image source). 2023. Available from: https://www.smart-elements.com/shop/activated-aluminum-oxide-spherical-3-2mm-2

[90] Desicca Chemicals. What is Alumina and what is it used for? (image source) 2023. Available from: https://medium.com/@desiccachemical4/what-is-alumina-and-what-is-it-used-for-f978d81acbd5

[91] Bararpour T, Karami D, Mahinpey N. Investigation of the effect of alumina-aerogel support on the CO_2 capture performance of K_2CO_3. Fuel. 2019;**242**:124-132. DOI: 10.1016/j.fuel.2018.12.123

[92] Bararpour T, Karami D, Mahinpey N. Utilization of mesoporous alumina-based supports synthesized by a surfactant-assisted technique for post-combustion CO_2 capture. Journal of Environmental Chemical Engineering. 2021;**9**(4):105661. DOI: 10.1016/j.jece.2021.105661

[93] Montagnaro F, Silvestre-Albero A, Silvestre-Albero J, Rodríguez-Reinoso F, Erto A, Lancia A, et al. Post-combustion CO_2 adsorption on activated carbons with different textural properties. Microporous and Mesoporous Materials. 2015;**209**(2015):157-e164. DOI: 10.1016/j.micromeso.2014.09.037

[94] Teng H, Yeh T, Hsu L. Preparation of activated carbon from bituminous coal with phosphoric acid activation. Carbon. 1998;**36**(9):1387-1395. DOI: 10.1016/S0008-6223(98)00127-4

[95] Sabzehmeidani M, Mahnaee S, Ghaedi M, Heidari H, Roy V. Carbon based materials: A review of adsorbents for inorganic and organic compounds. Materials Advances. 2021;**2**(2):598-627. DOI: 10.1039/D0MA00087F

[96] Wang T, Zhai Y, Zhu Y, Li C, Zeng G. A review of the hydrothermal carbonization of biomass waste for hydrochar formation: Process conditions, fundamentals, and physicochemical properties. Renewable and Sustainable Energy Reviews. 2018;**90**:223-247. DOI: 10.1016/j.rser.2018.03.071

[97] Sodimate. Activated Carbon: Types, Applications, Advantages. (image source). 2023. Available from: https://sodimate-inc.com/activated-carbon-types-applications-advantages

[98] Gao Y, Yue Q, Gao B, Li A. Insight into activated carbon from different kinds of chemical activating agents: A review. Science of the Total Environment. 2020;**746**:141094. DOI: 10.1016/j.scitotenv.2020.141094

[99] Parshetti G, Chowdhury S, Balasubramanian R. Biomass derived low-cost microporous adsorbents for efficient CO_2 capture. Fuel. 2015;**148**:246-254. DOI: 10.1016/j.fuel.2015.01.032

[100] Sahoo G, S. Experimental investigation on different activated carbons as adsorbents for CO_2 capture. Thermal Science and Engineering Progress. 2022;**33**:101339. DOI: 10.1016/j.tsep.2022.101339

[101] Wu D, Xu Q, Liu D, Zhong C. Exceptional CO_2 capture capability and molecular-level segregation in a Li-modified metal organic framework. Journal of Physical Chemistry C. 2010;**114**(39):16611-16617

[102] Silva P, Vilela S, Tomé J, Almeida P. Multifunctional metalorganic frameworks: From academia to industrial applications. Chemical Society Reviews.

2015;**44**:6774-6803. DOI: 10.1039/c5cs00307e (image source)

[103] Wen H, Liao C, Li L, Alsalme A, Alothman Z, Krishna R, et al. A metal organic framework with suitable pore size and dual functionalities for highly efficient post-combustion CO2 capture. Journal of Materials Chemistry A. 2019;**7**(7):3128-3134. DOI: 10.1039/C8TA11596F

[104] Qian D, Lei C, Hao G, Li W, Lu A. Synthesis of hierarchical porous carbon monoliths with incorporated metal organic frameworks for enhancing volumetric based CO_2 capture capability. ACS Applied Materials & Interfaces. 2012;**4**(11):6125-6132. DOI: 10.1021/am301772k

[105] Yu F, Phalak N, Sun Z, Fan L. Activation strategies for calcium-based sorbents for CO_2 capture: A perspective. Industrial and Engineering Chemistry Research. 2012;**51**(4):2133-2142. DOI: 10.1021/ie200802y

[106] Indiamart. Metal Oxide Nanoparticles (image source). 2023. Available from: https://www.indiamart.com/proddetail/metal-oxide-nanoparticle-15993165088.html

[107] Chowdhury S, Parshetti G, Balasubramanian R. Post-combustion CO_2 capture using mesoporous TiO_2/graphene oxide nanocomposites. Chemical Engineering Journal. 2015;**263**:374-384. DOI: 10.1016/j.cej.2014.11.037

[108] Liang P, Qin Y, Hu B, Peng T, Jiang Z. Nanometer-size titanium dioxide micro-column on-line pre-concentration of trace metals and their determination by inductively coupled plasma atomic emission spectrometry in water. Analytica Chimica Acta. 2001;**440**(2):207-213. DOI: 10.1016/S0003-2670(01)01010-8

[109] Li W, Wu Z, Wang J, Elzatahry A, Zhao D. A perspective on mesoporous TiO_2 materials. Chemistry of Materials. 2014;**26**(1):287-298. DOI: 10.1021/cm4014859

[110] Salaudeen S, Acharya B, Dutta A. CaO-based CO_2 sorbents: A review on screening, enhancement, cyclic stability, regeneration and kinetics modelling. Journal of CO2 Utilization. 2018;**23**:179-199. DOI: 10.1016/j.jcou.2017.11.012

[111] Li P, Chen R, Lin Y, Li W. General approach to facile synthesis of MgO-based porous ultrathin nanosheets enabling high-efficiency CO_2 capture. Chemical Engineering Journal. 2021;**404**:126459. DOI: 10.1016/j.cej.2020.126459

[112] Ruhaimi A, Aziz M, Jalil A. Magnesium oxide-based adsorbents for carbon dioxide capture: Current progress and future opportunities. Journal of CO2 Utilization. 2021;**43**:101357. DOI: 10.1016/j.jcou.2020.101357

[113] Mendoza E, Santos A, Lopez E, Drozd V, Durygin A, Chen J, et al. Iron oxides as efficient sorbents for CO_2 capture. Journal of Materials Research and Technology. 2019;**8**(3):2944-2956. DOI: 10.1016/j.jmrt.2019.05.002

[114] Nawar A, Ali M, Khoja H, Waqas A, Anwar M, Mahmood M. Enhanced CO_2 capture using organic acid structure modified waste eggshell derived CaO sorbent. Journal of Environmental Chemical Engineering. 2021;**9**:104871. DOI: 10.1016/j.jece.2020.104871

[115] Doonan C, Tranchemontagne D, Glover G, Hunt R, Yaghi M. Exceptional ammonia uptake by a covalent organic framework. Nature Chemistry. 2010;**2**(3):235-238. DOI: 10.1038/nchem.548

[116] Dawson R, Stockel E, Holst J, Adams D, Cooper A. Microporous organic polymers for carbon dioxide capture. Energy & Environmental Science. 2011;**4**(10):4239-4245. DOI: 10.1039/C1EE01971F

[117] Wang L, Xie W, Xu G, Zhang S, Yao C, Xu Y. Synthesis of thiophene-based conjugated microporous polymers for high iodine and carbon dioxide capture. Polymers for Advanced Technologies. 2022;**33**(2):584-590. DOI: 10.1002/pat.5540

[118] Dong B, Wang L, Zhao S, Ge R, Song X, Wang Y, et al. Immobilization of ionic liquids to covalent organic frameworks for catalyzing the formylation of amines with CO_2 and phenylsilane. Chemical Communications. 2016;**52**(44):7082-7085. DOI: 10.1039/C6CC03058K

[119] Penchah H, Najafi P, Ghaemi A, Gilani H. Characterization of hyper-crosslinked polymer adsorbent based on carbazole to achieve higher CO_2 capture. Environmental Progress & Sustainable Energy. 2021;**40**(4):13586-13586. DOI: 10.1002/ep.13586-13586

[120] Fayemiwo K, Chiarasumran N, Nabavi S, Loponov K, Manovic V, Benyahia B, et al. Eco-friendly fabrication of a highly selective amide-based polymer for CO_2 capture. Industrial and Engineering Chemistry Research. Washington, United States: American Chemical Society (ACS); 2019;**58**(39):18160-18167. DOI: 10.1021/acs.iecr.9b02347

[121] Haiyan Mao. et al. Menaline-based Tailpipe Carbon Capture Device (Online Resource). 2021. Available from: https://news.berkeley.edu/2022/08/04/a-simple-cheap-material-for-carbon-capture-perhaps-from-tailpipes [Accessed: January 16, 2023]

[122] Eller K, Henkes E, Rossbacher R, Höke H. Amines, aliphatic. In: Ullmann's Encyclopedia of Industrial Chemistry. DOI: 10.1002/14356007.a02_001

[123] Subramanian T, Sonthalia A, Varuvel E. Effect of calcite/activated carbon-based post-combustion CO_2 capture system in a biodiesel-fueled CI engine: An experimental study. Energy Sources, Part A: Recovery, Utilization and Environmental Effects. 2018;**2018**(41):1972-1982

[124] Thiyagarajan S, Edwin V, Martin L, Nagalingam B. Carbon dioxide (CO_2) capture and sequestration using biofuels and an exhaust catalytic carbon capture system in a single-cylinder CI engine: An experimental study. Biofuels. 2018;**9**:659-668

[125] Thiyagarajan S, Varuvel E, Martin L, Beddhannan N. Mitigation of carbon footprints through a blend of biofuels and oxygenates, combined with post-combustion capture system in a single cylinder CI engine. Renewable Energy. 2019;**130**:1067-1108

[126] Thiyagarajan S, Geo V, Martin L, Nagalingam B. Combined effect of fuel-design and after-treatment system on reduction of local and global emissions from CI engine. Environmental Technology. 2019;**40**:2802-2812

[127] Saravanan S, Kumar C. Carbon dioxide capture using adsorption technology in diesel engines. International Journal of Renewable Energy Research. Ankara, Turkey: Gazi University; 2020;**10**:1614-1620. DOI: 10.20508/ijrer.v10i4.11450.g8046

[128] Ajith-Kumar P, Varghese A, Vaishanth S, Balaji G. Development and test of a new carbon capture system using zeolite with addition of activated carbon and monoethanolamine for IC engines.

IOP Conference Series: Materials Science and Engineering. 2020;**2020**:912

[129] Kumar R. Experimental Investigations on CO_2 Recovery from Engine Exhaust Using Adsorption Technology. SAE Technical Paper; 2019

[130] Kaushal A, Ahirwar R, Vardhan A. Investigation of CO_2 capturing capacity of solid adsorbents (PEIs) polyethylenimines from automotive vehicle exhausts system for 4-stroke SI engine. Bonfring International Journal of Industrial Engineering and Management Science. 2018;**2018**(8):31-35

[131] Rajdurai M, Rao A, Kamalakkannan K. CO_2 capture using activated alumina in gasoline passenger vehicles. Fuel. 2016;**140**:4300RPM

[132] Saudi Aramco. Transport Technologies: Mobile Carbon Capture. 2023. Available from: https://www.aramco.com/en/what-we-do/energy-innovation/transportation/mobile-carbon-capture [Accessed: October 6]

[133] Ahmed S, Atilhan M. Evaluating the performance of a newly developed carbon capture device for mobile emission sources. Journal of Energy Resources Technology. 2017;**139**:1-8

[134] Remora. Heavy-duty fleet tests tailpipe CO_2 capture, compressor device (Online resource). 2023. Available from: https://www.www.remoracarbon.com

[135] Qaptis. CO_2 Capture Technology (Online resource). 2023. Available from: https://ggba.swiss/en/qaptis-mobile-co2-capture-device-revolutionizes-heavy-transport-decarbonization [Accessed: January 16]

Chapter 3

Carbon Reduction through Building Matrix Design for Sustainable Development

Bin Li, Luca Caneparo, Yuqing Zhang and Weihong Guo

Abstract

Under the influence of planet problems such as global warming and the energy crisis, carbon reduction has become one of the goals for sustainable development. This chapter focuses on residential buildings as research objects for achieving carbon reduction based on architectural design. After analyzing three design principles from the life cycle of the building, macro, medium and micro design scales, and passive-active design, the authors initially established an open-source matrix design method. Then, an architectural competition plan that won only first prize from 92 proposals was taken to explain further the connotation of carbon reduction to achieve sustainable development through the matrix design method. The results show that the vicious circle exists between global warming and the energy crisis. Residential buildings contain more potential than other buildings for solving these problems. The matrix design method is established based on the three design principles. From a qualitative perspective, according to the case application, this research helps achieve carbon reduction for sustainable development.

Keywords: matrix design, carbon reduction, sustainable development, residential building, architectural design, life cycle, passive design, renewable energy

1. Introduction

1.1 Background

At least from the nineteenth century, scientists began to recognize the role of carbon dioxide in global warming [1]. Using the NASA global temperature data from 1880 to 2022 [2], 'Tableau Public' (https://www.tableau.com/products/public) software was used to analyze and display the data. **Figure 1** shows the change in global surface temperature compared to the long-term average from 1951 to 1980. It can be clearly seen that the global land-ocean temperature index has been increasing in the past 150 years. On March 19, 2023, the Intergovernmental Panel on Climate Change (IPCC), the United Nations body for assessing the science related to climate change, published the Synthesis Report of the IPCC Sixth Assessment Report (AR6), which

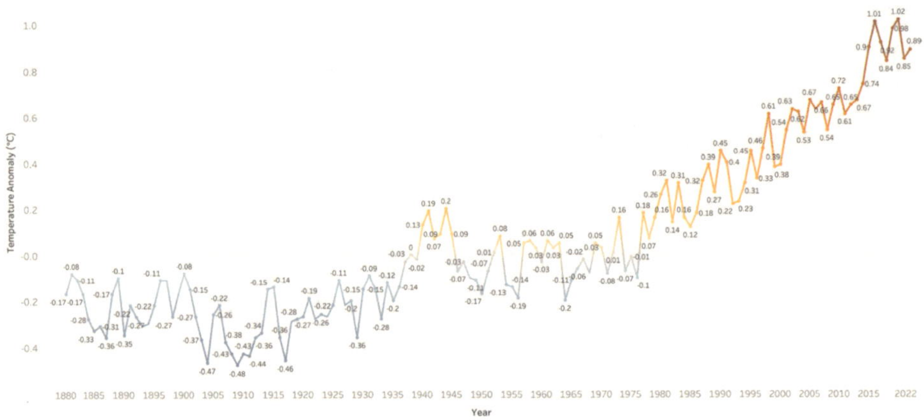

Figure 1.
Global land-ocean temperature index (the data come from Ref. [2], and the image was created by the authors).

indicates the global surface temperature reaching 1.1°C above 1850–1900 in 2011–2020 [3]. Global warming is intensifying and a problem that needs to be solved.

On the other hand, energy crises from fossil energy sources happen now and then. For non-renewable reasons, the more fossil energy is used, the less is left. Using the data downloaded from Our World in Data [4], **Figure 2** is generated by Tableau Public software to show global fossil fuel consumption. By 2021, especially since the Industrial Revolution, the consumption is increasing dramatically daily. Correspondingly, the remaining global fossil energy up to 2021 is shown in **Figure 3**. Although these values can change with time based on the discovery of new reserves and changes in annual production, the remaining years from 2020 of global coal, gas and oil are

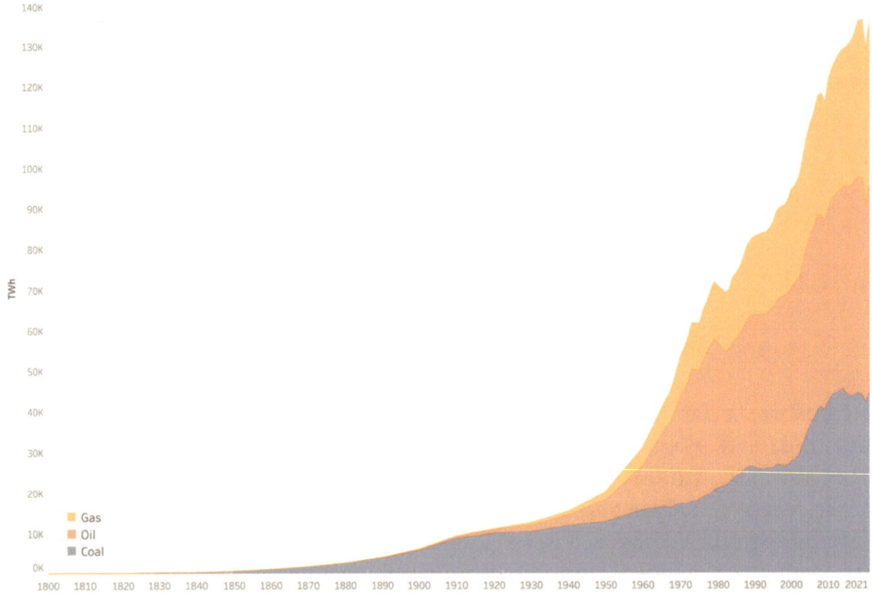

Figure 2.
Global fossil fuel consumption (the data come from Ref. [4], and the image was created by the authors).

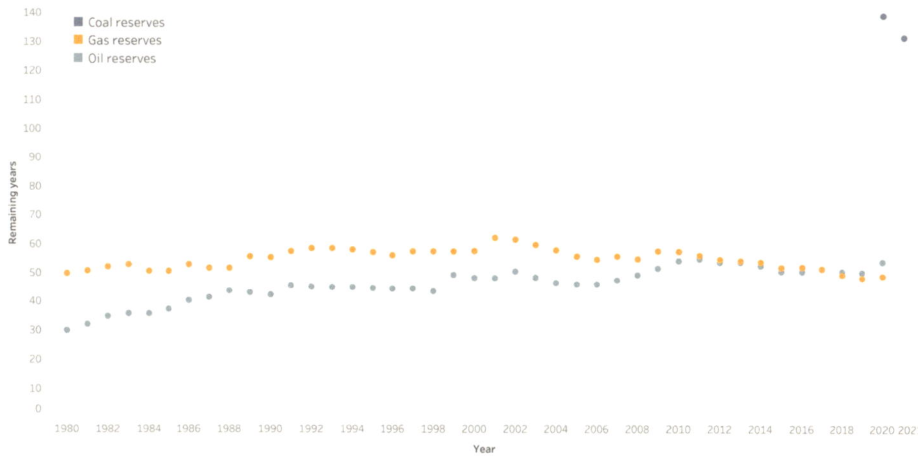

Figure 3.
Years of fossil fuel reserves left (the data come from Ref. [5], and the image was created by the authors).

only 139, 49 and 57, respectively [5]. Fossil energy consumption causes energy crisis and carbon emissions to exacerbate global warming [6]. A vicious circle is formed between global warming and the energy crisis. So, carbon reduction is essential to solving these problems for sustainable development.

1.2 Research objective

The 2022 Global Status Report for Buildings and Construction, published by the UN Environment Programme (UNEP), indicates that the buildings and construction sector accounted for around 37% of energy-and process-related CO_2 emissions and over 34% of energy demand globally in 2021. The building sector's operational energy-related CO_2 emissions reached an all-time high of around 10 $GtCO_2$, and operational energy demand in buildings reached an all-time high of 135EJ [7]. Moreover, residential buildings accounted for 17% of operational and process CO_2 emissions globally compared with 11% of non-residential buildings. The final energy demand for residential buildings is 21%, more than twice the proportion used in non-residential buildings (9%) [8]. So, energy saving, even producing renewable energy on residential buildings, is necessary and more effective for solving the energy crisis and global warming problems to some extent [9].

In this research, residential buildings are focused as research objects, especially from the building design aspects for achieving sustainable development by carbon reduction.

2. Design principles

The typical building design principles include the life cycle of the building in the time dimension, the macro, medium and micro design scales in the space dimension, and passive-active design in the technology dimension. These three design principles support each other to form the final buildings. The relationship between them is shown in **Figure 4**.

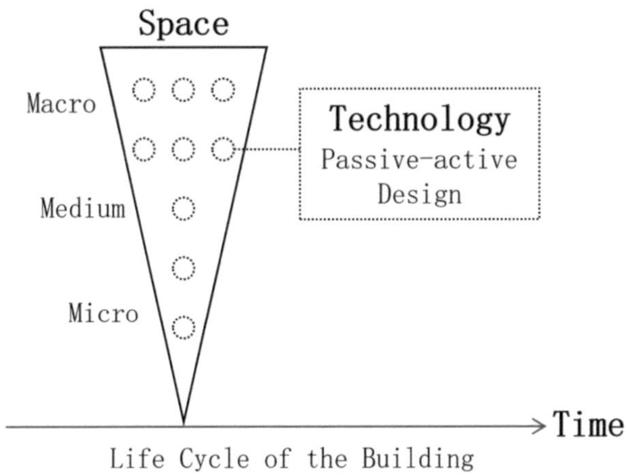

Figure 4.
The relationship of the three design principles.

2.1 Life cycle of the building

The life cycle of the building can be defined as five stages: design, construction, operation, renewal, and demolition [10]. The design and renewal, the construction and demolition, can be seen as similar and reversible stages, respectively, but they face different conditions (**Figure 5**).

The design stage is the point at which the most basic decisions regarding sustainable development, including carbon reduction, are taken [11]. It plays a critical role in determining the sustainability, the life cycle environmental impact and the efficiency of the buildings [12]. The construction stage can also take the role of carbon reduction but more from the construction practices, craft techniques and building materials which play a major role in the building's carbon footprint. The operation stage accounts for the most significant proportion of energy consumption in the building life cycle [13]. However, the user behavior and the energy efficiency of devices have a huge impact on that. The renewal stage can be regarded as redesign. Although it is not a design from scratch, optimizing the existing conditions with a better design is generally more sustainable [14]. The demolition stage can be seen as the inverse of the construction stage, especially if all the components can be reused.

All the stages can contribute to sustainable development with carbon reduction. However, the design and renewal stages play the leading role.

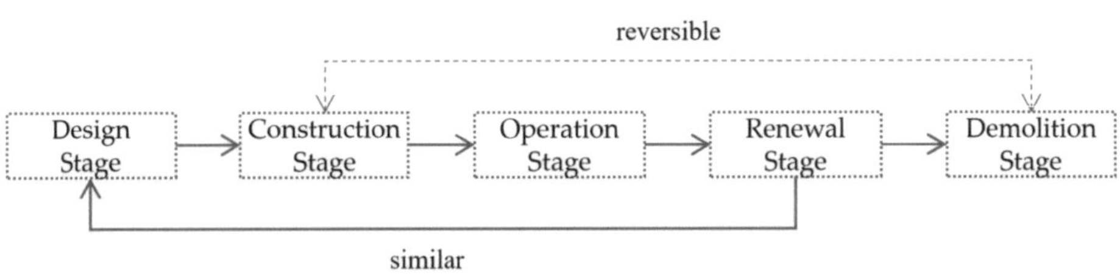

Figure 5.
The analysis of the life cycle of the building.

DOI: http://dx.doi.org/10.5772/intechopen.1004867

2.2 Macro, medium and micro design scales

In addition to the diachronic life cycle of the building, the different levels of synchronicity in the residential building are also essential design principles that should be paid attention to. The different levels of residential design can be divided into macro, medium and micro, matching community planning, building design and detail design [15], which makes up the second design principle.

Different levels of residential design focus on different scales. For example, the community planning of different residential buildings in Lingnan focuses mainly on site selection, layout models and related aspects for sustainable development (**Figure 6**).

1. Typical layout in Guangfu residential community.

2. Typical layout in Chaoshan residential community.

3. Typical layout in Hakka residential community.

As for the building design, it focuses more on space design. **Figure** 7 shows an example of a typical residential building design in Lingnan, highlighting the natural ventilation, to achieve carbon reduction. The design contents are different from the community planning.

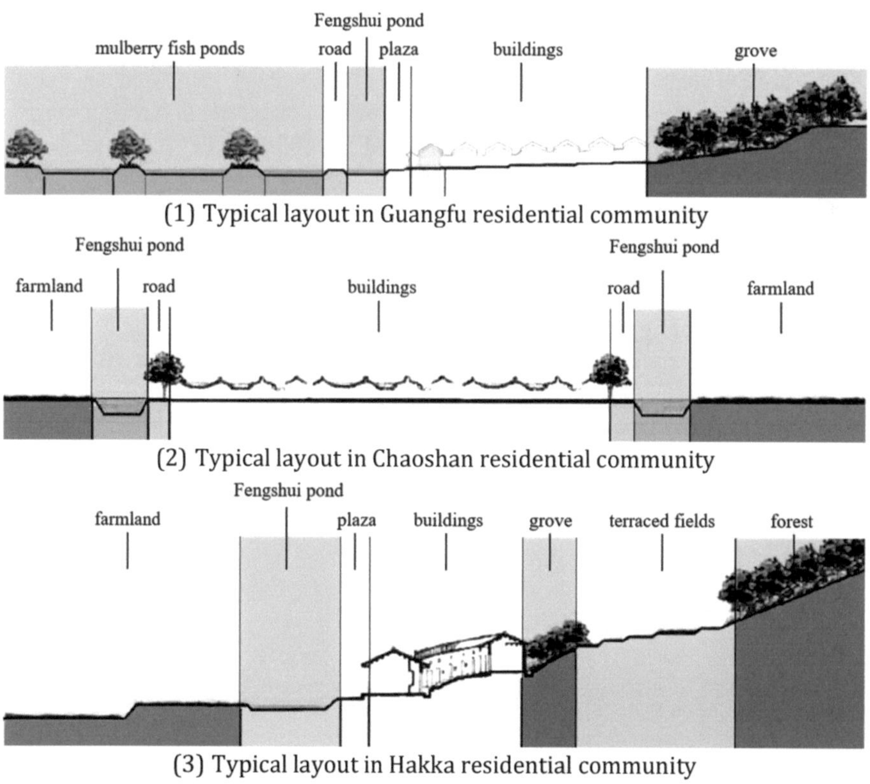

Figure 6.
Different community planning in Lingnan residential buildings.

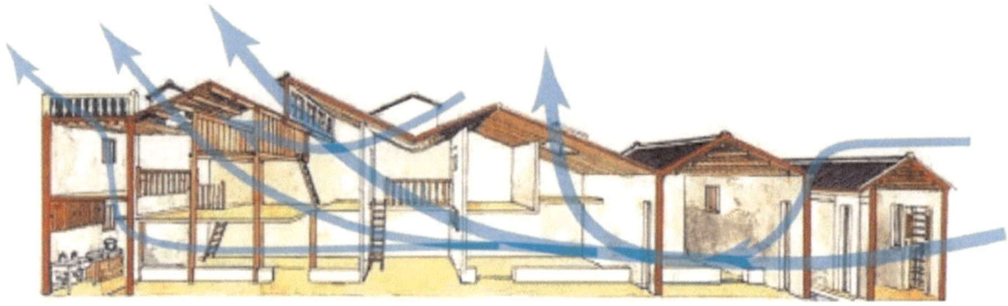

Figure 7.
Space design in Lingnan for natural ventilation.

At micro scale, detail design focuses more on the technologies that help to achieve community planning and building design. For example, the building technologies of the Oyster wall [16], different wall masonry methods and so on all support the implementation of residential buildings in Lingnan with different roles (**Figure 8**). These also contribute to achieving sustainable development.

2.3 Passive-active design

Generally, passive design in building is the leading way to reduce energy demand based on high performance. Active design can achieve carbon reduction by improving energy efficiency, such as using high-efficiency equipment. Even though some research indicates that active strategies include generating renewable energy, such as using solar panels [17], other scholars classified passive, active and renewable strategies as three factors for carbon reduction in buildings [18]. This research acknowledges active and passive classification methods. However, reducing energy demand, improving energy efficiency and generating renewable energy are different ways to reduce carbon.

Based on a 3rd Solar Decathlon China case study, 35 types of technologies are sorted out from the 15 competition residential buildings (**Figure 9**). It is clear that passive and active technologies, including reducing energy demand, improving energy efficiency, and generating renewable energy, are three aspects. These technologies are helpful for carbon reduction to achieve sustainable development.

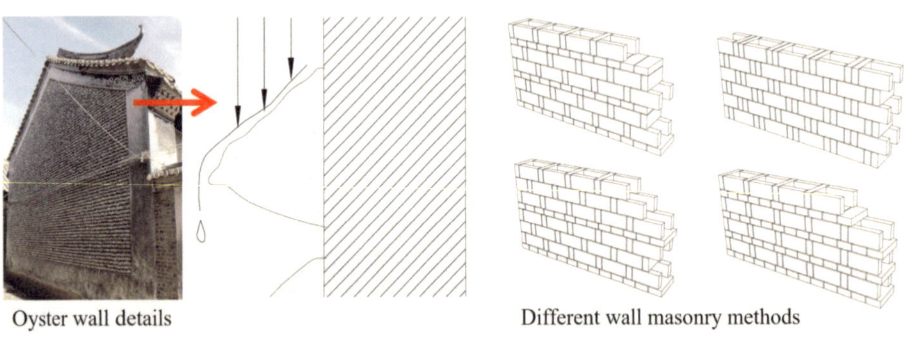

Figure 8.
Windows, doors and construction implementations in Lingnan.

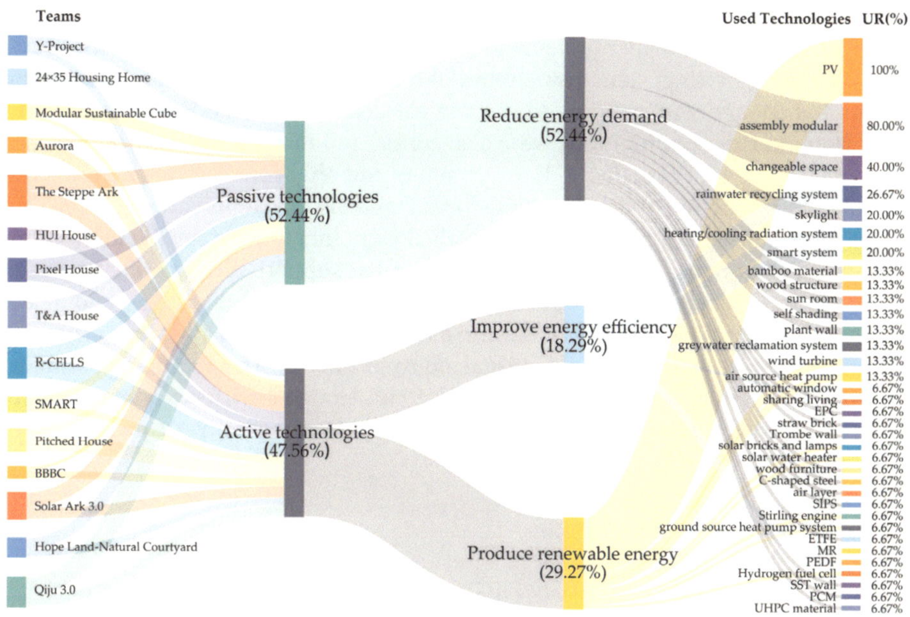

Figure 9.
Technologies used in the 3rd solar decathlon China (the data come from Ref. [19], and the image was created by the authors).

3. Matrix design method

3.1 Matrix design establishment

The matrix design method is grounded in three design principles. Although some methods exist to guide further design, these methods usually only follow some of these three design principles.

For example, the Life Cycle Assessment (LCA) method analyzes and quantifies the environmental consequences of buildings during their whole life cycle, from the extraction of raw materials, through production, use phase and end-of-life. This method is used to quantitatively assess the material used, energy flows and environmental impacts of products [20]. However, this method is more from the product point of view rather than architectural design. Moreover, a systematic literature review demonstrated that LCA continues to face barriers in both methods and practice, preventing its ability to guide early-stage design decisions and greatly impacting the environmental performance of buildings [21].

As for the design scales, these have split into different disciplines, such as urban design, architectural design, building construction and so forth. However, carbon reduction for sustainable development also needs knowledge of building technology, building services engineering (BSE) and further ones. This knowledge should be comprehensively applied to specific situations to achieve sustainable development [22].

Moreover, some research focus on the passive-active design. Aparicio-Ruiz et al. [23] did the research in Spain and tried to get the integrated optimization result for residential building design. The passive optimization design mainly considered the materials based on the best Energy Efficiency Index (EEI) and the lowest Life Cycle Cost (LCC). The annual energy demand for cooling and heating was calculated to define the energy rating. Generating renewable energy by solar panels was calculated to cover the energy consumption. Moreover, the active systems HVAC (heating, ventilation, and air conditioning) were also considered to maintain comfortable environments. However, these methods are complex and only consider carbon reduction from the passive-active design principle for sustainable development.

The matrix design method tries to find the key points for sustainable design based on the three design principles. A matrix diagram has supported the analysis. The matrix diagram is a project management and planning tool for displaying and analyzing the relationships between two or more data sets, also known as matrix charts. Matrix diagrams provide a valuable visual aid to intuitively understanding complex information that might otherwise be difficult to understand. Different types of matrix diagrams have been explored, such as L-shaped matrix diagram, Y-shaped, C-shaped, T-shaped and X-shaped.

3.2 Background on matrix design

Matrix design methods are rooted in "architecture, the age-old question of the quantification of quality, that is, of the measurability of the levels of satisfaction of the needs of living, answered by the performance approach. The demand-performance approach expresses the exceeding of the practice of associating specific 'objects' with specific 'needs' and shifts theoretical reflection and the possibilities of improving living conditions to the satisfaction of the latter. As a result, unlike traditional experiential approaches based on the rule of art, it is committed in 'what to get,' not 'how to get it'" [24].

Rational design theories started to be widely deployed in the mid-sixties as paradigms to ground design practices on "scientific methods" [25]. Christopher Alexander advanced demand-performance scientific methodologies, for instance, outlining design issues by means of detailed requirements and misfit variables [26]. Alexander is committed to bridging the gap between design practices and scientific methods: "This conflict, which has not yet been mended, has been repeatedly resumed, such as in the debate on the contrast between the two cultures, the humanistic and the scientific one, highlighted by Charles Snow, or in symmetric oppositions between world of ideas and world of practice. In architecture, in the wake of these dichotomies, there is still a contrast between a 'world of forms' and a 'world of techniques'" [27].

On the use of methodologies and techniques in design, Alexander [26] states "A design problem is not an optimization problem. In other words, it is not a problem of meeting any one requirement or any function of a number of requirements in the best possible way (though we may sometimes speak loosely as though it were, and may actually try to optimize one or two things like cost or construction time). For most requirements, it is important only to satisfy them at a level which suffices to prevent misfit between the form and the context, and to do this in the least arbitrary manner possible."

Giuseppe Ciribini outlines the methodology to represent the requirements in matrix form: the "Performance requirements (occurrences to be met through the production of a good) are translated into requirements:

- Environmental, issues or requests to the environment to fulfill suitable conditions for well-being, safety, etc.;

- Technological, or quality levels are required to technological objects or bodies to achieve the desired environmental quality.

A requirement (and the associated performance) can be expressed by one or more functions such as:

> f (a, b, c, ...),
> where a, b, c, ... are the characterizing variables. [...]

Their correlation in the system starts with filling a matrix with the correspondences between requirements, paired two by two" [28].

The Performance Based Building (PeBBu) Thematic Network has addressed a systematic approach for applying building industry and user requirements to the building process. Among the support toolkits, Design Structure Matrix (DSM) has been selected and trialed. DSM is described as "a compact and clear representation of complex system and capturing method for the interactions between system elements. Visual relationship matrix reveals key information flows and sets simultaneously targets to process analysis and re-engineering. It is used for finding the optimal order of tasks, defining product architecture (modularity and interfaces) and forming teams in large organisations. In case the problem exists it helps also to solve inconsistencies. There are many commercial applications available for DSM" [29].

3.3 Proposed matrix design methodology

Here, the L-shaped matrix diagram is used for analyzing three data sets. It is the intersection of the life cycle of the building data set with design scales, and passive-active design data set, respectively (**Table 1**).

3.4 Matrix design analysis

The life cycle of the building includes five stages. The interaction relationship between the life cycle of the building and design scales shows that only four stages

		Life cycle of the building				
		Design stage	**Construction stage**	**Operation stage**	**Renewal stage**	**Demolition stage**
Design scales	**Community planning**	√	√	×	√	√
	Building design	√	√	×	√	√
	Detail design	√	√	×	√	√
Passive-active design	**Reduce energy demand**	√	×	√	√	×
	Improve energy efficiency	√	×	√	√	×
	Generate renewable energy	√	×	√	√	×

Note: √ stands for relationship; and × stands for no relationship.

Table 1.
Matrix analysis for three design principles.

have relationships with community planning, building design and detail design. Even the operation stage can be considered at different scales, macro, intermediate, and micro (namely, the design scales: community planning, building design and detail design), the space was constructed before the operation stage. So, the operation stage depends more on user behavior and device energy efficiency, which are somewhat outside the building design. As for the construction and demolition stages, even though they are interactive with design scales, they point more to the construction modes, craft techniques and building materials on a broad design level.

As for the relationship between the life cycle of the building and passive-active design, there is no relationship between the construction and demolition stages because the passive, active and renewable aspects do not apply to the construction process. However, the passive, active and renewable aspects are important during the operation stage. As for their relationship in the design and renewal stages, these three aspects are essential to being involved in carbon reduction for sustainable development.

It can be found that design scales and passive-active design all take roles in the design and renewal stages. Referring to the Matrice d'impacts croisés multiplication appliquée à un classement (MICMAC) method, a cross-impact matrix multiplication [30], and the transitivity principle [31], the relationship between them can be established as a matrix method to sort out the key points for sustainable development.

3.5 Matrix design focus

The L-shaped matrix diagram is applied again, as **Table 2** shows. The intersecting sections show different key points for sustainable development in the design and renewal stages. Even though it is hard to sort out all the key points for matrix design, especially in the context of the continuous development of the times, new ideas and technologies will emerge. However, the intersection points of these three design principles were finally found through matrix analysis. These points are used to guide carbon reduction for sustainable development.

Community planning interacts with the three aspects of passive-active design, generating the layout, orientation, terrain, water, plant ..., pipe gallery, lighting, power, gas, water supply and drainage ... and energy station, biogas digester ..., respectively. Building design interacts with the three aspects of passive-active design,

		Passive-active design		
		Reduce energy demand	**Improve energy efficiency**	**Generate renewable energy**
Design scales	**Community planning**	Layout, orientation, terrain, water, plant ...	Pipe gallery, lighting, power, gas, water supply and drainage ...	Energy station, biogas digester ...
	Building design	Shape, space, color, window-to-wall ratio ...	HVAC, power, lighting, plumbing, smart control ...	PV, wind turbine, heat pump ...
	Detail design	Material, insulation, window, door, construction ...	Auto shading, sensor, craft ...	System, orientation, angle ...

Table 2.
Matrix design method.

generating the shape, space, color, window-to-wall ratio … , HVAC, power, lighting, plumbing, smart control … and PV, wind turbine, heat pump … , respectively. Detail design interacts with the three aspects of passive-active design, generating the material, insulation, window, door, construction … , auto shading, sensor, craft … and system, orientation, angle … , respectively. Some of them, such as improving energy efficiency in community planning and building design, generated similar results but different application scales. On the one hand, there may be only so many existing design points. On the other hand, some points may be repeated in different parts but face different problems.

Establishing the building matrix design method should be applied according to the specific conditions and developing with time as an open-source method. This method considers as many design points as possible from different design principles. Due to user's insufficient information, these may not be perfect enough. However, this can be regarded as a reorganization of the Architectural Practice System Theory for sustainable development [32]. Every design principle points to a system; its superposition is still a system but with different boundaries. Although the outline of the matrix design appears unsystematic and even fragmented, it can lead to problem-oriented outcomes and multi-objective solutions.

4. Case application

4.1 Introduction

Based on the background issues, the matrix design method is established after analyzing three design principles. The oriental courtyard design, designed by Bin Li et al. and which won first place out of 92 submitted designs [33], is taken as the case application of the building matrix design method for carbon reduction in residential buildings.

This competition requires the design of a residential building no more than 25 m in length and 18 m in width, a construction area of 240 $m^2 \pm 15$, a building height of less than 12 m, and not more than 3 floors. The building site should be in the Lingnan area [34]. The carbon reduction for sustainable development is one of the aims of this competition. It also encourages participants to learn from oriental architecture wisdom, traditional Chinese architectural culture, and then innovate to meet contemporary needs.

4.2 Design concept

The design site is chosen in Yongqingfang, Guangzhou city, where the density of bamboo barrel houses is high [35]. However, due to high density and long-term disrepair, the houses are dilapidated, resulting in huge energy demand that is not conducive to sustainable development. The residential buildings in Yongqingfang need to be renovated, even though there has now been a national renovation demonstration (**Figure 10**).

The design concept is based on building matrix design to figure out the renewal content. This project mainly focused on the building and detail design at the design scales. The passive-active design was applied together to reduce energy demand, improve energy efficiency and generate renewable energy. Therefore, the design points are mainly figured out in **Figure 11**. Moreover, the traditional Chinese

Figure 10.
Renovated Yongqingfang.

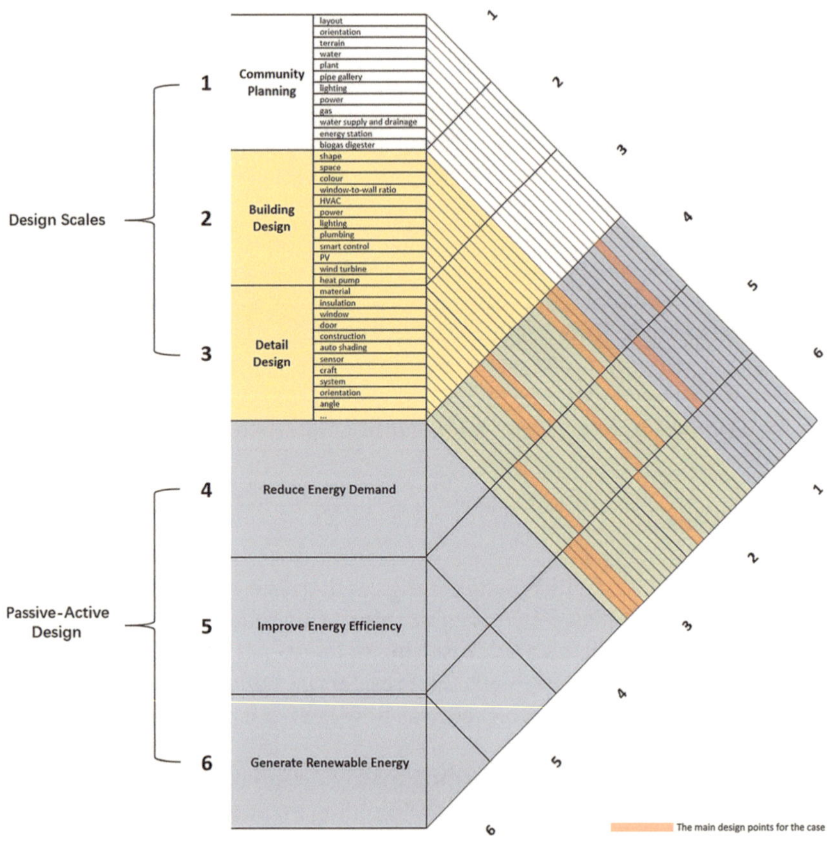

Figure 11.
The main design points of the renovation case (the figure was created by the authors).

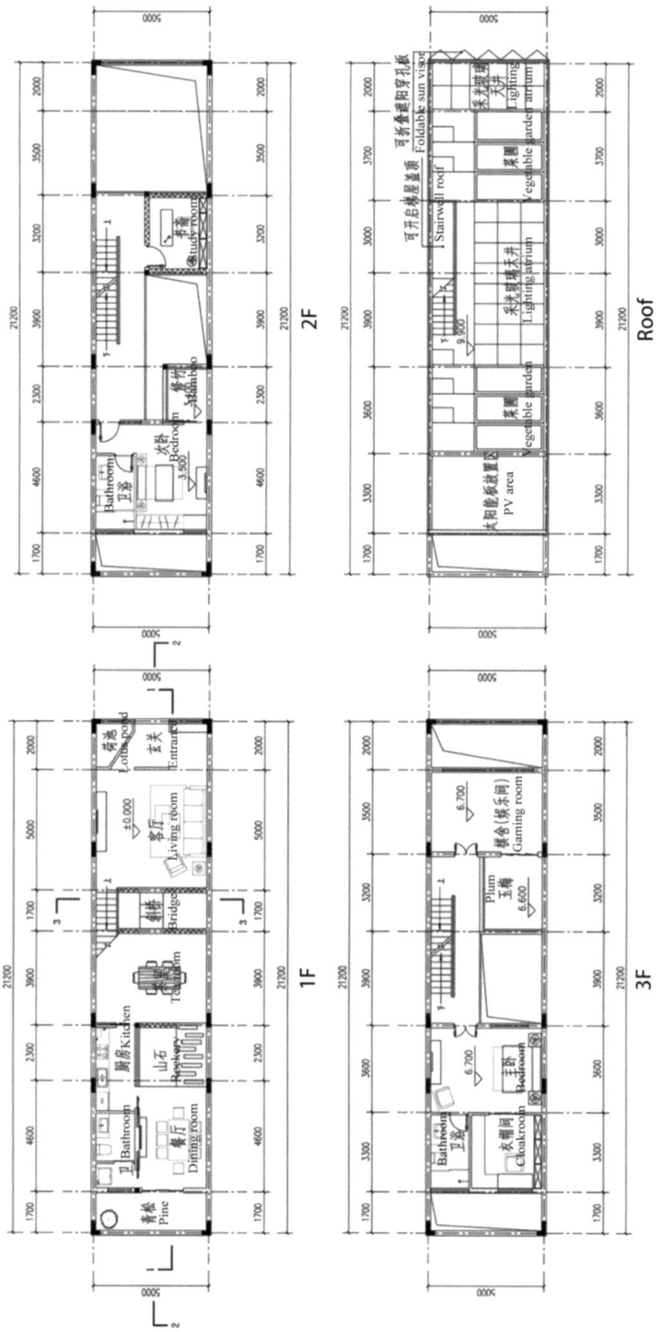

Figure 12.
Floor plan.

architectural culture is also considered according to the competition goals based on the open-source matrix design method. Yuyinshanfang, a traditional Lingnan Garden, was an attempt to learn for this design. The traditional passive technologies on the buildings and the greening measures were referenced.

4.3 Design results

The design-based bamboo barrel house is redesigned with three floors, including an accessible roof (**Figure 12**). The building function is shown on each floor. The sections indicate the building functions and the heights (**Figure 13**).

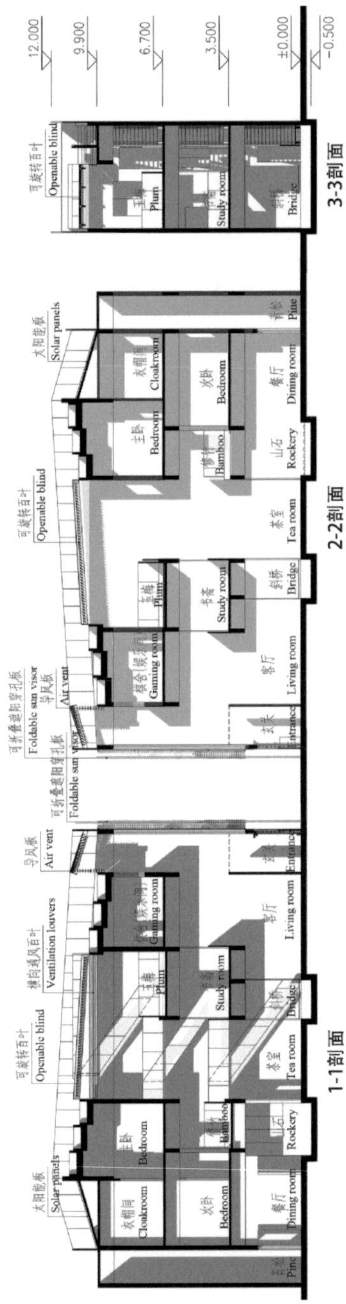

Figure 13.
Building section.

Traditional indoor split-level spaces are used for this design, which helps achieve wind-driven and buoyancy-driven ventilation (**Figure 14**). Phoenics software simulated the natural ventilation of this design (**Figure 15**) [36]. When the wind comes from the building entrance, the ventilation effect improves as floors rise. The wind flows out mainly from the central atrium and the rear courtyard, which are helpful to achieve a comfortable indoor environment without fossil energy consumption in the hot summer and warm winter Lingnan area and conducive to the discharge of the kitchen and bathroom smell. This helps to achieve carbon reduction by the passive building design for good indoor ventilation. The lighting environment is simulated by Ecotect (**Figure 16**). The results show that the overall indoor lighting is bright under the effect of atriums. However, in some places on the third floor, the lighting factor exceeds 7%, which may cause glare problems. Manual openable blinds can solve this without energy consumption. Passive lighting design helps conserve energy to reduce carbon emissions. The landscape elements and the plants used in the building for carbon reduction are shown in **Figure 17**. The above design points are passive design,

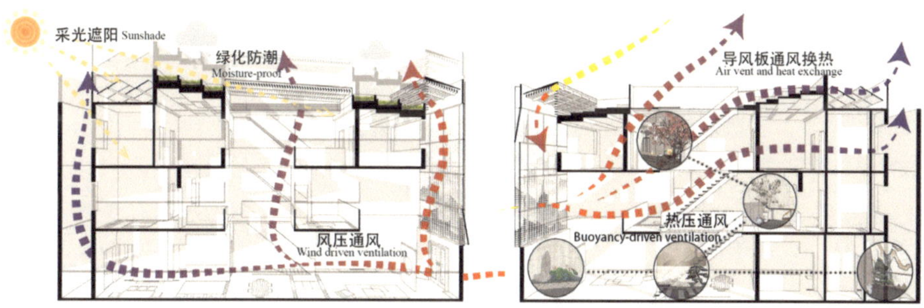

Figure 14.
Indoor ventilation.

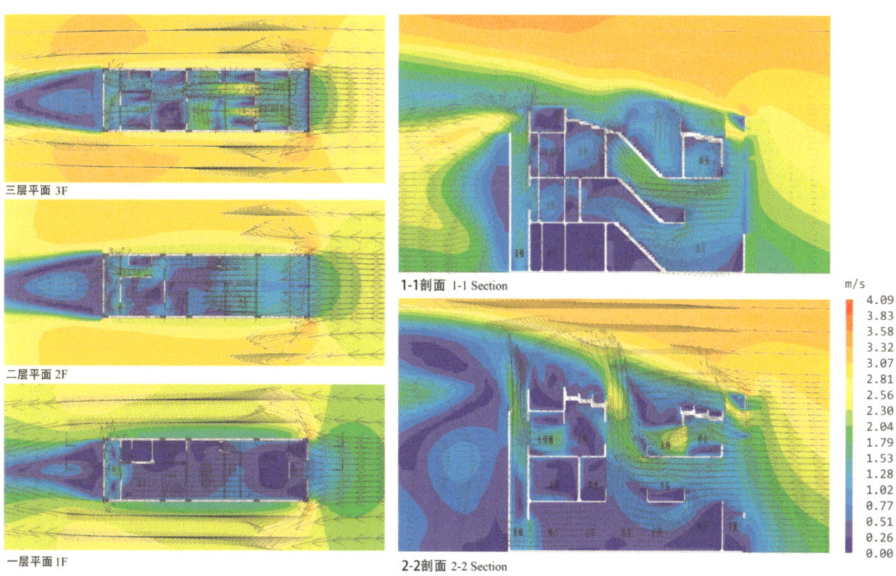

Figure 15.
Natural ventilation simulation.

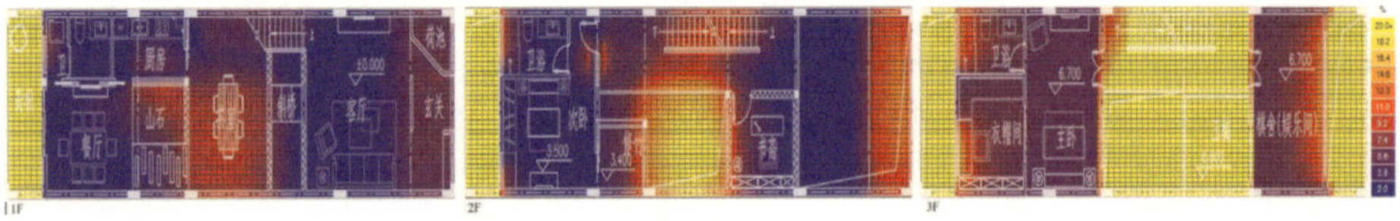

Figure 16.
Lighting environment simulation.

Figure 17.
Landscape in the building.

which avoids using energy through reasonable architectural design and achieves carbon reduction from the source. As for the active technologies, except for the smart control system for improving energy efficiency, the solar panels on the roof generate renewable energy to cover the building's energy consumption.

These key technologies include passive and active, as shown in **Figure 18**. Moreover, some interior renderings are shown in **Figure 19**. The design methods of matrix design and traditional Chinese architectural culture were applied to this project. The oriental courtyard spirit is inherited and innovated to meet the contemporary needs for carbon reduction to sustainable development.

5. Discussion and conclusions

This research starts from the global warming and energy crisis background problems to focus on the research object, residential buildings, for carbon reduction to achieve sustainable development. This problem-oriented research approach supports previous research results [37]. Based on the three design principles, the life cycle of the building, the macro, medium and micro design scales, and passive-active design, the relationship between them is figured out. That is, the integration of time, space

DOI: http://dx.doi.org/10.5772/intechopen.1004867

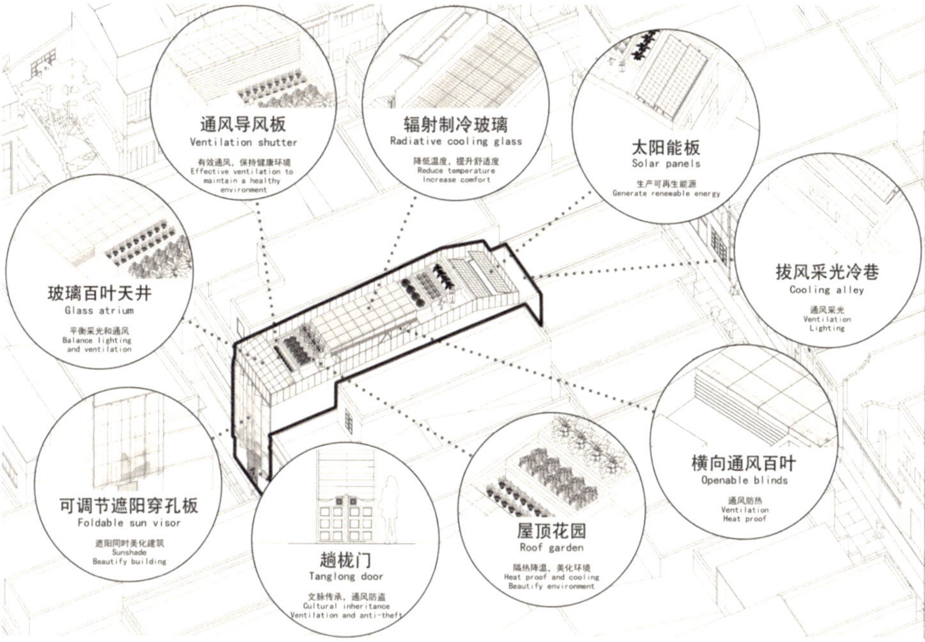

Figure 18.
Key technologies.

Figure 19.
Interior renderings.

and technology. The building matrix design method was originally established based on the design principles, which is an open-source design method that should be applied integrally according to the specific situation. Importantly, this method expands the architectural research ways [38, 39]. Then, the first prize of an architectural competition plan applied the matrix design method to show how to achieve carbon reduction in residential buildings for sustainable development.

Even though the building matrix design is established and applied in the competition plan, some limitations still need to be overcome. Due to the matrix design being an open-source method, it should be improved further. Moreover, the contributions of carbon reduction to sustainable development are hard to quantify for evaluation in this research, even though the direction helps achieve sustainable development [40]. This research mainly focuses on the original method established and applied for a case in a qualitative aspect. Further research from a quantitative aspect will be done in the future.

In summary, (1) the vicious circle relationship of background problems, global warming and energy crisis, is figured out. (2) Based on the problem-oriented, the most carbon reduction potential object for sustainable development in building, residential building, is identified. (3) The three design principles in architecture are analyzed, and the relationship is clarified. (4) A building matrix design method is originally advanced based on the design principles. This is an open-source method that expands the tools for architectural research. (5) Applying the matrix design method to a first-prize architectural competition plan illustrates how to reduce carbon in residential buildings for sustainable development.

Future research will focus on this direction to contribute more to sustainable development.

Acknowledgements

This research is supported by the Guangdong Basic and Applied Basic Research Foundation (Grant No. 2023A1515011137 and Grant No. 2024A1515012129); the 2022 Guangdong Philosophy and Social Science Foundation (Grant No. GD22XGL02); the Guangzhou Philosophy and Social Science Planning 2022 Annual Project (Grant No. 2022GZQN14); the Fundamental Research Funds for the Central Universities (Grant No. QNMS202211); the Guangzhou Basic and Applied Basic Research Foundation (Grant No. 2024A04J9930). Thanks to the MAECI Scholarship (Prot. N. 2793) for financial support. It is also partly supported by the China Scholarship Council (CSC) scholarship under the CSC Grant Nos. 202106150080 and 202006150053, and the Unite! Seed Fund project (Unite! Poster Contest on Architectural Energy for Sustainable Development).

Conflict of interest

The authors declare no conflict of interest.

DOI: http://dx.doi.org/10.5772/intechopen.1004867

Author details

Bin Li[1,2,3,4,5], Luca Caneparo[2*], Yuqing Zhang[1,2,5] and Weihong Guo[1,3,4,5]

1 School of Architecture, South China University of Technology, Guangzhou, China

2 Department of Architecture and Design, Politecnico di Torino, Turin, Italy

3 Energy Saving Technology Research Institute, South China University of Technology, Guangzhou, China

4 Architectural Design and Research Institute Co., Ltd., South China University of Technology, Guangzhou, China

5 State Key Laboratory of Subtropical Building Science, South China University of Technology, Guangzhou, China

*Address all correspondence to: luca.caneparo@polito.it

References

[1] Bonneuil C, Choquet P-L, Franta B. Early warnings and emerging accountability: Total's responses to global warming, 1971–2021. Global Environmental Change. 2021;**71**:102386

[2] NASA. Vital Signs [Internet]. 2023. Available from: https://climate.nasa.gov/vital-signs/global-temperature/ [Accessed: December 26, 2023]

[3] IPCC. Synthesis Report of the IPCC Sixth Assessment Report (AR6) [Internet]. 2023. Available from: https://report.ipcc.ch/ar6syr/pdf/IPCC_AR6_SYR_SPM.pdf [Accessed: December 26, 2023]

[4] Our World in Data. Fossil fuels [Internet]. 2023. Available from: https://ourworldindata.org/fossil-fuels#global-fossil-fuel-consumption [Accessed: December 26, 2023]

[5] Our World in Data. Years of fossil fuel reserves left, 2020 [Internet]. 2023. Available from: https://ourworldindata.org/grapher/years-of-fossil-fuel-reserves-left [Accessed: December 26, 2023]

[6] Dai H, Mamkhezri J, Arshed N, Javaid A, Salem S, Khan YA. Role of energy mix in determining climate change vulnerability in G7 countries. Sustainability. 2022;**14**(4):2161

[7] United Nations Environment Programme. 2022 Global Status Report for Buildings and Construction: Towards a Zero - emission, Efficient and Resilient Buildings and Construction Sector [Internet]. 2023. Available from: https://wedocs.unep.org/bitstream/handle/20.500.11822/41133/Building_Construction_2022.pdf?sequence=3&isAllowed=y [Accessed: December 26, 2023]

[8] United Nations Environment Programme. Tracking Progress [Internet]. 2023. Available from: https://globalabc.org/our-work/tracking-progress-global-status-report [Accessed: December 26, 2023]

[9] Papoyan A, Zhan C, Han X, Li G. Energy saving strategy for residential buildings: Case study in Armenia. International Journal of Low-Carbon Technologies. 2021;**16**(3):987-997

[10] Wong JKW, Zhou J. Enhancing environmental sustainability over building life cycles through green BIM: A review. Automation in Construction. 2015;**57**:156-165

[11] Azhar S, Carlton WA, Olsen D, Ahmad I. Building information modeling for sustainable design and LEED® rating analysis. Automation in Construction. 2011;**20**(2):217-224

[12] Kovacic I, Zoller V. Building life cycle optimization tools for early design phases. Energy. 2015;**92**:409-419

[13] Morsi DMA, Ismaeel WSE, Ehab A, Othman AAE. BIM-based life cycle assessment for different structural system scenarios of a residential building. Ain Shams Engineering Journal. 2022;**13**(6):9-13

[14] Zheng B, Masrabaye F, Guiradoumngué GM, Zheng J, Liu L. Progress in research on sustainable urban renewal since 2000: Library and visual analyses. Sustainability. 2021;**13**(8):1-4

[15] Zheng Y, Sun Y, Wang Z, Liang F. Developing green–building design strategies in the Yangtze River delta, China through a coupling relationship

between geomorphology and climate. Land. 2023;**12**(1):6

[16] Shuo W, Jian T, Shaobin W, Chenchen W, editors. Study on the unsteady thermal insulation performance of subtropical oyster shell wall constructions. In: Proceedings of the 2023 9th International Conference on Architectural, Civil and Hydraulic Engineering (ICACHE 2023). Dordrecht, The Netherlands: Atlantis Press; 2023

[17] Vahidinasab V, Ardalan C, Mohammadi-Ivatloo B, Giaouris D, Walker SL. Active building as an energy system: Concept, challenges, and outlook. IEEE. 2021;**9**:58009-58024

[18] Kim C-H, Park M-K, Kang W-H. Energy saving quantitative analysis of passive, active, and renewable technologies in different climate zones. Applied Sciences. 2021;**11**(15):1-4

[19] Li B, Guo W, Liu X, Zhang Y, Caneparo L. The third Solar Decathlon China buildings for achieving carbon neutrality. Buildings. 2022;**12**(8):7-20

[20] Blengini GA, Di Carlo T. The changing role of life cycle phases, subsystems and materials in the LCA of low energy buildings. Energy and Buildings. 2010;**42**(6):869-880

[21] Roberts M, Allen S, Coley D. Life cycle assessment in the building design process – A systematic literature review. Building and Environment. 2020;**185**: 107274

[22] Leoto R, Lizarralde G. Challenges in evaluating strategies for reducing a building's environmental impact through integrated design. Building and Environment. 2019;**155**:34-46

[23] Aparicio-Ruiz P, Guadix-Martín J, Barbadilla-Martín E, Muñuzuri-Sanz J. Applying renewable energy technologies in an integrated optimization method for residential building's design. Applied Sciences. 2019;**9**(3):1-10

[24] Laurìa, A. Approccio esigenziale-prestazionale e qualità dell'abitare. Claudi de Saint Mihiel, A.(ed.) Tecnologia e progetto per la ricerca in Architettura, Napoli: Clean Edizioni; 2014; 54.

[25] Blachère G, Bâtir S. Habitabilité, durabilité, économie des bâtiments, Eyrolles. Paris: Eyrolles; 1965

[26] Alexander C. Notes of the Synthesis of Form. Cambridge, MA: Harvard University Press; 1964

[27] Losasso M. Between theories and practices: Culture, technology. TECHNE-Journal of Technology for Architecture and Environment. 2017; **13**:9

[28] Ciribini G. Tecnologia e progetto: argomenti di cultura tecnologica della progettazione. Turin: Celid; 1984

[29] Porkka J, Huovila P. Decision Support Toolkit (DST) – A step towards an Integrated Platform for Performance Based Building (PBB). In: Huovila P, editor. Combining Forces - Advancing Facilities Management & Construction through Innovation Series: Performance Based Building. Vol. V. Helsinki, Finland: VTT Technical Research Centre of Finland; 2005. pp. 81-93

[30] Ahmad M, Tang X-W, Qiu J-N, Ahmad F. Interpretive structural modeling and MICMAC analysis for identifying and benchmarking significant factors of seismic soil liquefaction. Applied Sciences. 2019;**9**(2):1-13

[31] Dubey R, Ali SS. Identification of flexible manufacturing system

dimensions and their interrelationship using total interpretive structural modelling and fuzzy MICMAC analysis. Global Journal of Flexible Systems Management. 2014;**15**(2):131-143

[32] Guo W. Architectural Practice System Theory. Guangzhou: South China University of Technology Press; 2016

[33] 筑岸设计. "东方人居杯"方圆东方的院子建筑设计竞赛入围作品 [Internet]. 2023. Available from: https://mp.weixin.qq.com/s/fJPUF7dkuy3NOIYi2vuoYA [Accessed: December 31, 2023]

[34] 南方建筑. 东方人居杯 [Internet]. 2023. Available from: https://mp.weixin.qq.com/s/QYl3UCvyqfHmp0RCyCUgWg [Accessed: December 31, 2023]

[35] RIBA. Bamboo Barrel House Museum in Guangzhou [Internet]. 2023. Available from: http://www.presidentsmedals.com/Entry-13671 [Accessed: December 31, 2023]

[36] Li B, Guo W, Schnabel MA, Moleta T. Feng-shui and computational fluid dynamics (CFD): Analyzing natural ventilation and human comfort. In: 25th International Conference on Computer-Aided Architectural Design Research in Asia, CAADRIA 2020, August 5, 2020–August 6, 2020. Bangkok, Thailand: The Association for Computer-Aided Architectural Design Research in Asia (CAADRIA); 2020

[37] Li B, Guo W, Liu X, Zhang Y, Russell PJ, Schnabel MA. Sustainable passive design for building performance of healthy built environment in the Lingnan area. Sustainability. 2021;**13** (16):1-3

[38] Zhang Y, Li B, Caneparo L, Meng Q, Guo W, Liu X. Physical environment study on social housing stock in Italian western Alps for healthy and sustainable communities. Land. 2023;**12**(7):11-24

[39] Zhang Y, Liu X, Meng Q, Li B, Caneparo L. Physical environment research of the family ward for a healthy residential environment. Frontiers in Public Health. 2022,**10**

[40] Li B, Zhang Y, Caneparo L, Guo W, Meng Q. Energy use in residential buildings for sustainable development: The fifth Solar Decathlon Europe revelations. Heliyon. 2024;**e30701**:30-34. DOI: 10.1016/j.heliyon.2024.e30701

Chapter 4

Impact of Rice Husk Biochar on Soil Carbon Sequestration, Methane Emission, and Rice Yield in Wetland Soil (Ultisol)

Kamala Rajalekshmi, Betty Bastin and Sandeep Sasidharan

Abstract

The integration of physical, chemical, and biological indicators of soil quality that have an impact on environmental factors and farmer profitability is the focus of soil health. Applying biochar is an effective technique to promote soil carbon sequestration, which is a component of soil quality. Rice husk biochar (RHB), produced from the pyrolysis of rice husk (RH), has a higher nutritious value than biochar made from wood. A field research was conducted to evaluate the short-term implications of RHB-derived nutrient translocation and distribution in rice crop, as well as their effects on soil nutrients, rice productivity, and methane (CH_4) emissions in wetland Ultisol. The treatments included applying four organic fertilizers: rice husk biochar, daincha (*Sesbania aculeata*), jack tree (*Artocarpus heterophyllus*) leaves, and farm yard manure (FYM) at rates of 35 (N1), 70 (N2), and 105 (N3) kg N ha^{-1}. The results showed that the application of RHB enhanced the uptake of soil organic carbon (SOC) as well as the production of rice grains and straw. The yield of brown rice increased with higher rates of N treatment. The RHB contributed a significant amount of carbon to the refractory paddy soil, which increased its carbon content in the crop as well. Consequently, there was a significant 50–60% drop in CH_4 emissions from soil as compared to FYM.

Keywords: biochar, soil characteristics, organic sources, yield of rice, methane emission

1. Introduction

Wetland paddy fields contribute significantly to greenhouse gas emissions in the present climate change scenario, particularly methane (CH_4) and carbon dioxide (CO_2), depending on the availability of water. The wetland rice fields are responsible for around 25% of the 500 Tg yr^{-1} of methane emissions that occur globally each year. Sequestration of soil organic carbon (SOC) is one of the most effective methods to increase soil organic matter (SOM) reserves and to mitigate the potential greenhouse effect [1]. The process of soil carbon sequestration is impacted by greenhouse gas emissions, which lower the amount of carbon stored in the soil. The types of soils, temperatures, types of plants, erosion, and land management all have a significant impact on the addition and removal of organic carbon from soils. Breakdown of SOC in cultivated soils has contributed approximately 50 Pg C to the atmosphere [2]. This has promoted a strategy of enhancing soil carbon sequestration in agricultural systems to offset CO_2 increase. The use of various organic sources, such as manures, plant leftovers, and waste materials, in farming might enhance the chemical and biological qualities of soils [3]. Several studies have reported that biochar application increased crop yield by improving the physicochemical properties [4, 5], whereas some studies have observed a reduction in grain yield due to N limitation caused by the high C:N ratio of biochar [6].

Biochar is a solid material created through the thermochemical conversion of biomass under oxygen-limiting circumstances. Biochar contains carbon (C) that remains in the soil for long periods of time, making it an excellent alternative for increasing soil carbon sequestration. However, the moderating effect of biochar application on global warming remains debatable since it may alter the production and consumption of greenhouse gases (GHGs) other than CO_2, such as methane (CH_4) and nitrous oxide (N_2O) in soil. Its effect on GHG emission varies according to circumstances such as soil type, fertility, and moisture content, as well as the feedstock and pyrolysis temperature utilized to produce the biochar [7]. Soil temperature and soil moisture are important factors controlling CO_2 fluxes by their influence on organic matter decomposition.

Rice husk biochar (RHB) is considered to be one of the most cost-effective biochars used in rice-based farming systems [8]. A by-product of rice production, rice husk makes up 22% of the yield of brown rice. When applied to soil, it has the potential to improve soil fertility, increase the mean residence time of soil organic C (SOC) content, and dramatically reduce greenhouse gas emissions [9–11].

Perennial cropping methods have recently acquired popularity because they reduce the escape of carbon into the atmosphere, aiding soil carbon storage and climate change mitigation efforts. In the tropics, however, short-term interventions may produce bigger responses because turnover durations for fast-cycling soil C are shorter [12].

Little attention has been made to the short-term impact of various organic sources on carbon sequestration and its fractions. In light of this, the current study was conducted in a rice crop under rainfed conditions to assess the effect of four contrasting organic sources, giving importance to the rice husk biochar, on carbon sequestration/emission and their effects on rice productivity in a typical Ultisol.

2. Materials and methods

2.1 Experimental design

A field experiment was conducted in a farmer's field near Agricultural University, Thrissur, Kerala, India (10°31′49″N and 76°12′53″E, 10 m above sea level) from September 2015 to February 2016 (**Figure 1**). The soil of the experimental site was sandy clay loam (Ultisol) with pH (4.66), available N (401.76 kg N ha^{-1}), and organic C content (1.05%). The experiment was laid out in a randomized block design with 5.0 m × 4.0 m plot size with three replications. The treatments included four organic

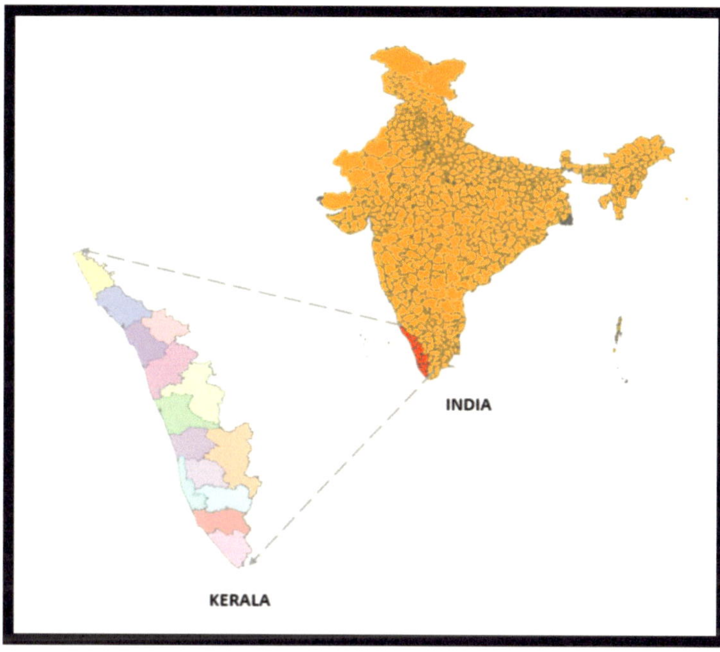

Figure 1.
Location of field.

sources [farm yard manure (FYM), *Artocarpus* (jack tree) leaves (ART), *Sesbania aculeata* (daincha) (DNC), and rice husk biochar (RHB)] in the absence or presence of N mineral fertilizer and the details are provided in **Table 1**. RHB was produced by pyrolysis at 350–400°C from the rice husk collected from a farmer's field near the experimental site. The physicochemical characteristics of the organic sources (FYM, DNC, ART, and RHB) are presented in **Table 2**.

Eighteen-day-old rice seedlings of short-duration variety 'Jyothi' were transplanted manually at a distance of 15 cm from row to row and 10 cm from plant to plant (**Figures 2** and **3**). CH_4 fluxes during rice cultivation period and soil chemical properties after rice harvest were measured. Yield and yield attributes were recorded, and straw, grain, and root samples were analyzed for nutrient content.

2.2 Crop management

Organic sources (FYM, DNC, ART, and RHB) were thoroughly mixed in the treatment plots 15 days prior to transplanting. Two rice seedlings (*Oryza sativa* L.) at the 3–4 leaf stage were transplanted as one hill. Urea, superphosphate, and muriate of potash were used to apply N, P, and K, respectively. Half of the N and K doses, as well as the entire amount of P, were applied as basal, with the remaining N and K administered as top dressing 1 week before panicle initiation. The daily mean air temperature ranged from 25.0°C to 31.0°C during the experiment period.

2.3 Measurements

The CH_4 flux was measured in triplicate following a closed chamber method [13]. A translucent acrylic chamber of 1.0 meters in height and 0.30 by 0.45 meters in bottom

Sl. no.	Treatments	Description
1	T_1	No organic manure + 35 kg N ha^{-1}
2	T_2	No organic manure + 70 kg N ha^{-1}
3	T_3	No organic manure + 105 kg N ha^{-1}
4	T_4	Farmyard manure @ 5 t ha^{-1} + 35 kg N ha^{-1}
5	T_5	Farmyard manure @ 5 t ha^{-1} + 70 kg N ha^{-1}
6	T_6	Farmyard manure @ 5 t ha^{-1} + 105 kg N ha^{-1}
7	T_7	*Artocarpus* sp. @ 5 t ha^{-1} + 35 kg N ha^{-1}
8	T_8	*Artocarpus* sp. @ 5 t ha^{-1} + 70 kg N ha^{-1}
9	T_9	*Artocarpus* sp. @ 5 t ha^{-1} + 105 kg N ha^{-1}
10	T_{10}	Daincha @ 5 t ha^{-1} + 35 kg N ha^{-1}
11	T_{11}	Daincha @ 5 t ha^{-1} + 70 kg N ha^{-1}
12	T_{12}	Daincha @ 5 t ha^{-1} + 105 kg N ha^{-1}
13	T_{13}	Rice husk biochar @ 5 t ha^{-1} + 35 kg N ha^{-1}
14	T_{14}	Rice husk biochar @ 5 t ha^{-1} + 70 kg N ha^{-1}
15	T_{15}	Rice husk biochar @ 5 t ha^{-1} + 105 kg N ha^{-1}

Table 1.
Treatment details.

Sl. no.	Property	Rice husk biochar	Farmyard manure	*Artocarpus heterophyllus* (leaves)	Daincha
1	pH	9.10	7.5	4.85	5.80
2	EC (dS m^{-1})	0.26	0.31	0.14	0.45
3	OC (%)	43.7	22.32	20.36	9.17
4	C:N	68.28	12.90	27.83	11.70

Table 2.
Physicochemical properties of organic sources.

area was placed over six hills, and it was sealed with water surrounding it (**Figure 4**). Gas samples were collected at three stages of the crop, *viz.*, active tillering, panicle initiation, and near harvest. The emission of gases from the plots was compared to that from the fallow wetlands. At the start of the sampling (time, t = 0), air sample from the box was drawn into a 20 ml plastic syringe fitted with a three-way stop-cock and closed. A second sample was collected in the same manner after a time interval of 20 minutes (time, t = 20 min). Sampling was done between 11.00 AM and 01.00 PM. The samples were brought to the laboratory and methane concentrations were determined by gas chromatography on a Flame Ionization Detector (FID). The CH_4 flux was calculated as a temporal change in the gas concentration at 0 and 20 min. The soil redox potential (Eh) at a depth of 5 cm during the experiment was monitored using a portable Eh meter. Cumulative CH_4 flux and its effect on yield as yield-scaled CH_4 flux were also calculated.

The soil samples collected after the rice harvest were sieved (<2 mm) and analyzed. Soil pH and electrical conductivity (soil:H_2O @ 1:2.5) were determined using

Figure 2.
Field preparation.

Figure 3.
View of transplanted paddy field.

a pH meter and EC meter. The rice grain and straw yield and yield parameters like root biomass, LAI, productive tillers per panicle, percentage of ripened grains, and 1000 grain weight were also recorded. Grain, straw, and root samples were separately analyzed for total N content. The N uptakes were calculated by multiplying the nutrient concentrations and the rice grain or straw weight for each plant.

2.4 Statistical analyses

Analysis of variance (ANOVA) was done using OP-STAT and WASP to test the level of significant difference between treatment means [14].

Figure 4.
Box chamber with hills placed in the field for GHG collection.

2.5 Growth, yield, and yield components of rice crop

Among the yield parameters, root biomass and LAI increased with higher N rates (**Table 3**). The maximum values were obtained in treatments having daincha with 70 kg N ha^{-1} (137.80 g m^{-2} and 6.77) and FYM with 70 kg N ha^{-1} (137.01 g m^{-2} and 6.61). The number of productive tillers per panicle and percentage of ripened grains were also maximum in FYM + 70 kg N ha^{-1} (6.50 and 82.60) followed by the treatment-receiving daincha +70 kg N ha^{-1} (6.39 and 81.31). The lowest value for the yield attributes was measured in treatment with no organic manure. A maximum grain yield of 3940.33 kg ha^{-1} was recorded with the treatment having daincha and 35 kg N ha^{-1} while RHB with nitrogen at 105 kg ha^{-1} produced a comparable grain yield of 3732.13 kg ha^{-1} along with a maximum straw yield of 4879 kg ha^{-1}. Liming effect of biochar in acidic soil along with increased nutrient retention may be the most important factors for increased crop yields [15]. Further, the 'fertilizer effect' of biochar with the supply of important plant nutrients such as K, N, Ca, and P in soil could have an additional effect on crop yield. With RHB at higher N application rates, the number of productive tillers per panicle and the percentage of ripened grains increased which had a positive effect on brown rice yield and the straw yield while 1000 grain weight showed a reverse trend. The depletion of SOC status and related nutrients may be the cause of the low grain yield of rice in the treatment that did not receive any organic manure. These outcomes concur with the research reported in Kumari et al. [16].

2.6 N concentrations in rice plants

The maximum grain N (**Figure 5**) was obtained with FYM + 70 kg N ha^{-1} (1.29%), and daincha at the same dose of N recorded at 1.28%. FYM, with a higher level of N retained 1.26%. The maximum straw nitrogen (0.65%) was observed in treatment

Treatment	Yield (kg ha^{-1})	Root biomass (g m^{-2})	LAI tillers	No. of productive filled grains	LAI tillers	Percentage of weight (g)	1000 grain
	Grain	Straw					
T_1	3489.00	4722.33	128.67	6.54	5.16	73.82	23.4
T_2	3529.33	4606.00	129.43	6.58	5.69	73.72	23.69
T_3	3562.33	4582.00	132.02	6.60	5.29	74.56	24.36
T_4	3771.67	4374.67	135.46	6.57	5.57	80.21	25.21
T_5	3894.00	4477.33	137.01	6.61	6.50	82.60	28.10
T_6	3831.67	4294.33	136.67	6.59	5.94	82.06	27.27
T_7	3536.00	4360.00	129.70	6.57	5.18	74.45	24.53
T_8	3576.67	4431.67	131.69	6.58	5.50	75.83	25.47
T_9	3662.33	4590.00	132.60	6.57	5.63	76.33	25.78
T_{10}	3940.33	4416.67	135.75	6.75	6.37	80.19	26.83
T_{11}	3866.67	4496.00	137.80	6.77	6.39	81.31	27.17
T_{12}	3813.33	4445.67	136.62	6.73	5.66	77.88	26.7
T_{13}	3687.33	4651.67	129.62	6.59	5.33	74.69	25.62
T_{14}	3726.73	4689.33	130.83	6.62	5.67	75.95	25.53
T_{15}	3732.13	4879.00	133.15	6.64	5.68	76.09	25.58
CD (0.05)	28.61	135.14	0.54	0.014	0.39	1.64	0.73

Table 3.
Growth, yield and yield components of rice crop.

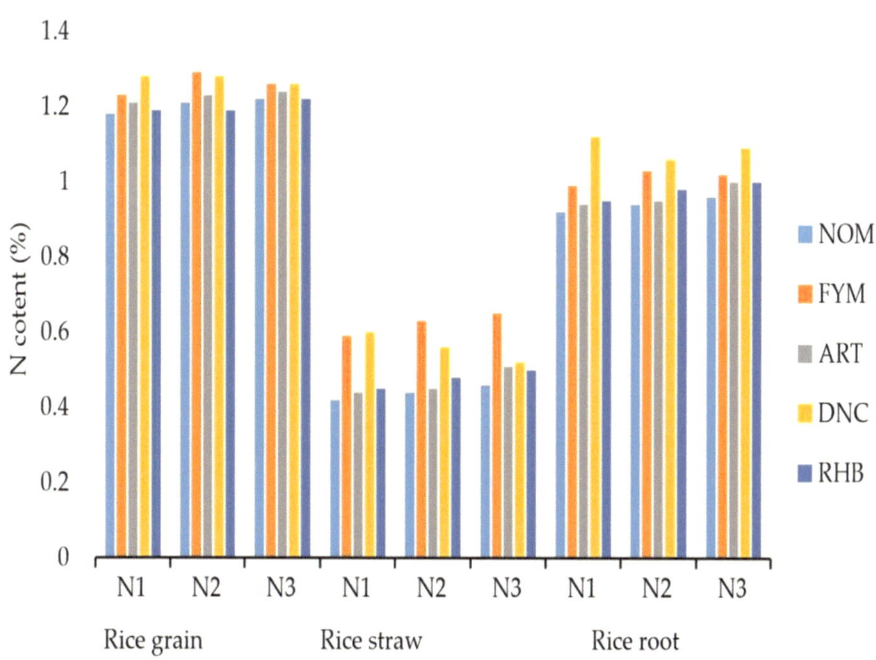

Figure 5.
N concentrations of rice crop.

with FYM + 105 kg N ha^{-1}. Treatments, daincha, and FYM with 35 kg N ha^{-1} recorded on par values. Daincha with 35 kg N ha^{-1} had maximum N content in the root (1.12%) while it was 1.09% in treatment-receiving daincha with 105 kg N ha^{-1}.

The N content was found to be higher in plant parts when the organic sources like FYM and daincha were amended with a medium dose of nitrogen (70 kg N ha^{-1} as per KAU POP) than higher levels of N. This could be ascribed to the inhibition of SOM decomposition by adding urea at higher levels, which necessitates additional C requirements for microbes. A higher value of N content in grain, root, and straw could also be attributed to the ability of organic manures to supply nutrients continuously during the growth period of crop as a result of mineralization and improvement of the physical and chemical properties of the soil. This result is in line with the findings of Sharma et al. [17]. Many scientists have also observed negative or non-significant effects of N on the decomposition of organic matter [18, 19].

2.7 Soil chemical properties and C sequestration

The maximum pH of 5.15 was recorded for the treatment having RHB with a lower N level (**Table 4**) owing to the alkalinity of RHB (pH 9.1) and liming value of biochar [20] while the treatment with 105 kg nitrogen without any organic manure had a minimum pH of 4.58. According to most reports, the presence of ash residues including significant levels of silica, heavy metals, sesquioxides, phosphates, and trace quantities of organic and inorganic nitrogen was responsible for the preferred pH shift brought about by the application of biochar [21].

Treatment	pH	EC	OC	Avail. N
T_1	4.60	0.045	0.64	427.85
T_2	4.59	0.056	0.94	438.23
T_3	4.58	0.058	0.82	451.49
T_4	63	0.055	1.16	463.35
T_5	4.62	0.055	1.14	474.30
T_6	4.62	0.064	1.1	459.40
T_7	4.63	0.043	1.20	425.89
T_8	4.61	0.046	1.01	431.61
T_9	4.61	0.048	1.01	453.98
T_{10}	4.73	0.063	1.06	590.69
T_{11}	4.70	0.117	1.14	595.71
T12	4.70	0.145	1.12	596.38
T_{13}	5.15	0.045	1.94	475.44
T_{14}	5.01	0.086	1.23	477.28
T_{15}	5.00	0.144	1.05	487.51
CD AxB (0.05)	0.012	0.001	0.20	1.637

Table 4.
Chemical properties of soil.

When compared to other organic sources, the RHB treatment had a greater SOC concentration; this was greatest in the RHB + 35 kg N (1.94%) treatment. This 50–70% increase above the control [22] might be the result of applying C-rich organic material (RHB) to acidic soils, which raised the pH of the soil and its SOC content [23]. Further, the low rate of degradation of organic matter in waterlogged conditions compared to the upland aerobic situations could have favored SOC addition. Similar results were obtained by Rajalekshmi and Bastin [10].

Significant variation in EC of the soil was noticed and it ranged between 0.043 dS m^{-1} and 0.145 dS m^{-1} in the top 15 cm of the soil. However, the individual effect of organic sources or nitrogen levels on EC was non-significant. The highest EC value of 0.145 dS m^{-1} was observed in treatment having daincha and 105 kg N ha^{-1} while minimum EC (0.043 dS m^{-1}) was associated with ART. RHB with a lower dose of N recorded a value of EC (0.045 dS m^{-1}) and with increasing doses of N, the EC touched 0.144 dS m^{-1} which was on par with the highest value showing the impact of an additional dose of inorganic N.

Available N was the highest where the application of daincha (*Sesbania aculeata*) was adopted along with fertilizer application and was maximum in the treatment having daincha +105 kg N ha^{-1} (596.38 kg ha^{-1}). It was also observed that the soil available nitrogen increased with increasing N levels of nitrogen, but with SOC content, the reverse trend was noticed. Ding et al. [15] found that green manuring combined with fertilizer application increased the available N content in soil and our result was also the same. The extent of availability of nutrients depends not only on the type of organic additives added but also on the build-up of microorganisms-enhanced mineralization (**Figure 6**) [24]. The availability of soil N remained higher in the treatments with biochar (475.44–487.51 kg ha^{-1}) despite greater nutrient removal for plant growth and higher grain yields.

Figure 6.
Biochar-treated plot with earthworms.

2.8 Soil redox potential (Eh) and CH4 emission

During the paddy crop growth period, soil Eh steadily dropped and stayed at around −170 mV between 50 DAT. There was no discernible variation in soil Eh between the treatments during the trial. CH_4 fluxes increased in nearly all treatments when the crop reached the heading stage, or around 70 days after planting, and the soil Eh fell below −150 mV.

2.9 Methane emission

Methane emission at the active tillering stage (AT) was highest in the treatment having FYM with 35 kg N ha^{-1} (15.30 mg m^{-2} h^{-1}), and it was lowest (0.86 mg m^{-2} h^{-1}) with the same organic source at 105 kg N ha^{-1} (**Table 5**). The emission from the fallow plot was 1.36 mg m^{-2} h^{-1}. The emission rate of methane was in the order of FYM > daincha > no organic manure > *Artocarpus* > RHB.

The treatment, FYM with 35 kg N ha^{-1} had the highest methane emission (19.89 mg m^{-2} h^{-1}), and it was lowest in daincha +105 kg N ha^{-1} (0.11 mg m^{-2} h^{-1}) at the panicle initiation stage (PI). The rate of emission decreased when the quantity of N fertilizer increased to 105 kg N ha^{-1} with and without organic sources. Treatments with *Artocarpus*/RHB had emissions ranging from 10 to 15 mg m^{-2} h^{-1} while the fallow plot had an emission value of 2.97 mg m^{-2} h^{-1}.

Treatments	Methane emission (mg/m^2/h)			Cumulative methane emission (mg/m^2/h)	Yield scaled methane emission ($\times10^{-3}$) g CH_4/kg rice yield
	Active tillering	Panicle initiation	Near harvest		
T_1	7.47	10.40	2.21	20.08	57.55
T_2	5.76	13.38	0.14	19.28	54.63
T_3	6.46	4.88	0.12	11.46	32.17
T_4	15.30	19.89	0.15	35.34	93.70
T_5	12.86	17.32	0.94	31.12	79.92
T_6	0.86	1.23	0.13	2.22	5.79
T_7	6.63	15.19	0.13	21.95	62.10
T_8	8.41	13.01	0.87	22.29	62.32
T_9	3.95	12.10	0.14	16.19	44.21
T_{10}	9.57	12.03	1.10	22.70	57.61
T_{11}	5.06	18.21	0.16	23.43	60.61
T12	4.43	0.11	0.12	4.66	12.22
T_{13}	3.74	13.77	0.18	17.69	47.82
T_{14}	3.68	12.10	1.15	16.93	44.94
T_{15}	3.63	10.36	0.23	14.22	37.51
CD (AxB) (0.05)	0.016	0.017	0.017	0.018	0.013

Table 5.
Influence of organic sources on CH_4 emission.

At the near harvest stage (NH), maximum emission was noticed in the treatment with 35 kg N ha^{-1} (2.21 mg m^{-2} h^{-1}), and it was minimum with daincha + 105 kg N ha^{-1} (0.12 mg m^{-2} h^{-1}). RHB with 70 kg N ha^{-1} had the highest emission to the tune of 1.15 mg m^{-2} h^{-1} followed by daincha with 35 kg N ha^{-1} (1.1 mg m^{-2} h^{-1}). When the treatments were amended with 70 kg N ha^{-1}, the rate of emission increased except in daincha and inorganic N alone treatments. As the quantity of N was further increased, the emission value was reduced. The emission from the fallow plot was 0.14 mg m^{-2} h^{-1}.

Methane production was the highest in FYM-treated plots at active tillering and panicle initiation stages compared to near harvest stage. This may be attributed to the higher activity of microbial populations, especially methanogens in FYM-supplied soil. The addition of inorganic fertilizer might have acted as a source of electron donors, thus increasing methanogens in these treatments under a flooded rice soil system. Thus, urea could enhance CH_4 emissions over the growing seasons which might be due to an increase in soil pH following urea hydrolysis and a decrease in redox potential. These two processes could have enhanced methanogenic activities. Dubey [25] pointed out that CH_4 emission depends on the nature, quantity, and method of fertilizer application. Maximum emission was seen in the study by Linquist et al. [26].

2.10 CH4 fluxes

The treatment with FYM had significantly higher CH_4 flux than other organic sources and was maximum in FYM + 35 kg N ha^{-1} (35.34 mg m^{-2} h^{-1}) and minimum with 105 kg N ha^{-1} (2.22 mg m^{-2} h^{-1}). The increase in different stages was not proportional to the levels of inorganic sources. The cumulative CH_4 emission increased in FYM-treated plots by 55 to 75% to that of control (**Table 5**). RHB treatment recorded higher CH_4 flux from the AT to PI stage but it remained in lower levels at NH stage. Similarly, RHB with 105 kg N ha^{-1} recorded a minimum cumulative emission of 14.22 mg m^{-2} h^{-1} compared to other N levels. Consequently, the cumulative CH_4 emission for RHB was 50–60% lower than that for FYM, although differences between the treatments were non-significant.

It is evident that the physicochemical properties of the microbial population and the soil environment have an impact on the GHG emissions from organic sources. The cumulative methane emission was higher in FYM-treated plots compared to other organic sources and with biochar-amended soil, it was less. The use of daincha as an organic source significantly reduced both leaching and gaseous loss of fertilizer N in soil, and this is in line with the findings of Bhattacharyya and Mandal [27]. Similarly, due to the biological stabilization of carbon and nitrogen in soil, RHB reduced CH_4 emissions and, consequently the global warming potential (GWP).

2.11 Yield-scaled CH4 emission and soil C sequestration

The treatment-receiving RHB induced minimum CH_4 fluxes compared to other organic sources. The yield-scaled CH_4 emission increased with FYM by 30.00–45.00% compared to control T1 (**Table 5**), and a minimum was also recorded in this treatment (5.79×10^{-3} g/kg rice). In the RHB treatment, the CH_4 emission was reduced to nearly half compared to the FYM ($40–48 \times 10^{-3}$ g/kg rice), and it was less than control also. The results under yield and yield attributes of rice clearly showed significant variation among the treatments. The data showed that for producing the same kg of rice, the CH_4 emission was less in the RHB compared to other organic sources, and it followed the order FYM > ART > DNC > RHB. The values for DNC and RHB were on par. The growth attributes, leaf area index, and root biomass production had significant effect on yield. The organic amendments application resulted in increased plant growth and a higher leaf area index. This might be associated with the fact that a considerable amount of nutrients, especially nitrogen, could be released from organic materials for plant use, essential for chlorophyll and protoplasm formation. Further, due to the cementing action of polysaccharides and other organic compounds released during the decomposition of organic matter, taller and heavier plants were produced. Babu et al. [28] reported that plant height was significantly influenced by the application of organic manure and chemical fertilizers. These observations are in line with the findings of Mansour [29] and Selvaraj et al. [30]. Agronomic practices such as appropriate nutrient supply could increase the total amount of both root and shoot biomass [31]. Application of green manure or other organic fertilizers improved soil physicochemical and microbiological parameters, and they could have contributed to sustainable productivity in flooded rice soil. Further, the enhanced soil biological properties provided resiliency and buffering capacity of soil to ameliorate the stress [32].

3. Conclusion

The study showed that applying RHB to paddy soil, Ultisol increased both rice output and the short-term soil carbon budget. The paddy soil received notable levels of C from RHB. RHB supplied significant amounts of C to the paddy soil with minimal effect on CH_4 emission, thereby sequestering more carbon in soil. The cumulative methane emission was 50–60% less compared to the FYM in the wetland paddy soil, which ultimately reduced the yield-scaled emission. The uptake of N by rice plants increased with higher rates of inorganic source application. This led to a notable rise in the number of grains and percentage of matured grains in the rice plants, which in turn raised the production of rice. Application of RHB as organic source also yielded an appreciable rice yield. Since the nutrients in these organic sources are released gradually, RHB application in the field (equal to 5 Mg ha^{-1} of rice husk) could be recommended once over a period of 3 or 4 years, or in lesser amounts in ongoing applications. Long-term field studies are needed to determine the optimum RHB application rates and frequencies for increasing the rice yield under different soil types and environmental conditions.

Author details

Kamala Rajalekshmi*, Betty Bastin and Sandeep Sasidharan
Department of Soil Science and Agricultural Chemistry, Kerala Agricultural University, Vellanikkara, Thrissur, Kerala, India

*Address all correspondence to: rajalekshmiharish@gmail.com

References

[1] Lal R. Soil carbon sequestration to mitigate climate change. Geoderma. 2004;**123**(1):1-22

[2] Paustian K, Collins HP, Paul EA. Management controls on soil carbon. In: Paul EA, Paustian KA, Elliot ET, Cole CV, editors. Soil Organic Matter in Temperate Agroecosystems: Long-Term Experiments in North America. New York: CRC Press; 1997. pp. 15-49

[3] Goyal S, Mundra MC, Chander K, Kapoor KK. Influence of inorganic fertilizers and organic amendments on soil organic matter and soil microbial properties under tropical conditions. Biology and Fertility of Soils. 1999;**29**(2):196-200

[4] Glaser B, Lehmann J, Zech J. Ameliorating physical and chemical properties of highly weathered soils in the tropics with charcoal – A review. Biology and Fertility of Soils. 2002;**35**:219-230

[5] Lehmann J, Rondon M. Biochar soil management on highly eathered soil in the humid tropics. In: Uphoff N et al., editors. Biological Approaches to Sustainable Soil Systems. Boca Raton, FL: CRC Press; 2006. pp. 517-530

[6] Asai H, Benjamin KS, Stephan MH. Biochar amendment techniques for upland rice production in northern Laos: 1. Soil physical properties, leaf SPAD and grain yield. Field Crops Research. 2009;**111**:81-84

[7] Lehmann J, Joseph S. Biochar for environmental management: An introduction. In: Lehmann J, Joseph S, editors. Biochar for Environmental Management-Science and Technology. London: Earthscan Publisher; 2009. pp. 1-12

[8] Ogawa M, Okimori Y. Pioneering works in biochar research, Japan. Australian Journal of Soil Research. 2010;**48**(7):489-500

[9] Lehmann J, Cheng CH, Nguyen B, Liang B, Major J, Smernik R. Permanency and long term changes of bio-char in soil. In: International Agriculture Initiative (IAI) Conference, Terrigal, Australia. Biogeochemistry. 2007. p. 23

[10] Rajalekshmi K, Bastin B. Biochar as an organic source for soil carbon sequestration in acidic soil of Kerala. Journal of Soil and Water Conservation. 2019;**18**(3):307-310

[11] Rodríguez L, Salazar P, Preston TR. Effect of biochar and biodigester effluent on growth of maize in acid soils. Livestock Research for Rural Development. 2009;**21**:102-110

[12] Trumbore SE. Potential responses of soil organic carbon to global environmental change. Proceedings of the National Academy of Sciences of the United States of America. 1997;**94**:8284-8291

[13] Hutchinson GL, Mosier AR. Improved soil cover method for field measurement of nitrous oxide fluxes. Soil Science Society of America Journal. 1981;**45**(2):311-316

[14] Sheoran OP, Tonk DS, Kaushik LS, Hasija RC, Pannu RS. Statistical software package for agricultural research workers. In: Recent Advances in Information Theory, Statistics and Computer Applications, Department of Mathematics Statistics. Hisar: CCS HAU; 1998. pp. 139-143

[15] Ding Y, Liu Y, Wu W, Shi D, Yang M, Zhong Z. Evaluation of biochar effects on

nitrogen retention and leaching in multi-layered soil columns. Water, Air, and Soil Pollution. 2010;**213**:47-55

[16] Kumari GSK, Thakur SK, Kumar V, Kumar N, Singh SK. Long term effect of fertilizer, farm- yard manure and lime on yield sustainability and soil organic carbon pools under soybean (*Glycine max*)- wheat (*Triticum aestivum*) cropping system in Alfisol. Journal of Soil and Water Conservation. 2019;**18**(2):196-204

[17] Sharma MP, Bali SV, Gupta DK. Crop yield and properties of inceptisol as influenced by residue management under rice- wheat cropping sequence. Journal. Indian Society of Soil Science. 2000;**48**:506-509

[18] Ramirez KS, Craine JM, Fierer N. Nitrogen fertilization inhibits soil microbial organic carbon in deep soil layers controlled by fresh carbon supply: Nature nutrients. Netherlands Journal of Agricultural Science. 2010;**41**:327-341

[19] Vallack HW, Leronni V, Metcalfe DB, Hogberg P, Ineson P, Subke J. Effect of somotic, alkaline, acid or thermal stresses on the growth and inhibition of *listeria monocytogenes*. Journal of Applied Microbiology. 2012;**86**:469-476

[20] Arocena JM, Opio C. Prescribed fire-induced changes in properties of sub-boreal forest soils. Geoderma. 2003;**113**:1-16

[21] Raison RJ. Modification of the soil environment by vegetation fires, with particular reference to nitrogen transformation: A review. Plant and Soil. 1979;**51**:73-108

[22] KAU. Package of Practices Recommendations. Thrissur: Directorate of Extension, Kerala Agricultural University; 2011. 140 p

[23] Haynes RJ, Mokolobate MS. Amelioration of Al toxicity and P deficiency in acid soils by additions of organic residues: A critical review of the phenomenon and the mechanisms involved. Nutrient Cycling in Agroecosystems. 2001;**59**:47-63

[24] Kaur M, Saini SP. Effect of long term rice straw incorporation on soil phosphorus availability in a sandy loam soil. Journal of Soil and Water Conservation. 2019;**18**(2):144-149

[25] Dubey SK. Microbial ecology of methane emission in rice agroecosystem: A review. Applied Ecology and Environmental Research. 2005;**3**:1-27

[26] Linquist BA, Borbe MA, Pittelkow CM, Kessel CV, Groenigen KJV. Fertilizer management practices and greenhouse gas emissions from rice systems: A quantitative review and analysis. Field Crops Research. 2012;**135**:10-21

[27] Bhattacharyya K, Mandal SR. Effect of green manuring on growth parameters and yield of rainfed rice. Environment and Ecology. 1996;**14**:446-450

[28] Babu S, Marimuthu R, Manivannan V, Kumar RS. Effect of organic and inorganic manures on growth and yield of rice. Agricultural Science Digest. 2001;**21**(4):232-234

[29] Mansour HA. Influence of soil amendments and NPK fertilization on the growth and chemical composition of *Senna sulfurea* plants grown in a sandy soil. Annals of Agricultural Science. 2002;**40**(3):1779-1796

[30] Selvaraj N, Ramaraj B, Devrajan K, Sreenivasan N, Senthil Kumar S, Sakthi E. Effect of organic farming on growth and yield of thyme. In: National Seminar on Production

and Utilisation of Medicinal Plants. Tamilnadu: Annamalai University; 2003. pp. 13-14

[31] Grechi I, Vivin P, Hilbert G, Milin S, Robert T, Gaudille'̀re, J.P. Effect of light and nitrogen supply on internal C:N balance and control of root to-shoot biomass allocation in grapevine. Environmental and Experimental Botany. 2007;**59**:139-149

[32] Karlen DL, Eash NS, Unger PW. Soil and crop management effects on soil quality indicators. American Journal of Alternative Agriculture. 1992;7(1-2):48-55

Chapter 5

A Review of Strategies to Achieve Net Zero Targets in the Cement and Concrete Sectors

Kwaku Boakye, Dahl Winters, Olurotimi Oguntola, Kevin Fenton and Steve Simske

Abstract

The cement and concrete industries face significant challenges in decarbonization due to escalating demand driven by rapid urbanization, population growth, and infrastructure restoration needs. Cement production alone accounts for 8% of global anthropogenic CO_2 emissions, underscoring the urgency of exploring pathways to achieve net-zero emissions in these sectors. With over 120 nations committing to net-zero targets by 2050, a comprehensive examination of emerging carbon-saving technologies is imperative. While several promising innovations are in nascent stages, rigorous life cycle assessments are essential to determine their potential for carbon reduction. Practical strategies for achieving net-zero objectives and UN sustainability goals involve embracing circular economy principles, harnessing diverse by-product sources, fostering stakeholder engagement, and fostering technological innovation. An efficiency approach that integrates advancements in materials science, alternative fuels, and sector-wide efficiencies is crucial for success. This assessment identifies promising technologies, highlights knowledge gaps, underscores the importance of further research, and offers recommendations for implementing best practices on the path to net zero.

Keywords: cement, concrete, net zero, CO_2 emissions, decarbonization, climate change, sustainability

1. Introduction

There is global agreement that the biggest threat to the environment and the economy in our time is global warming. Research by Mahlia [1] and Zhang et al. [2] indicates that greenhouse gas (GHG) emissions associated with human activity cause global warming, which can have disastrous consequences if left unchecked. The frequency of extreme natural occurrences such as hurricanes, droughts, heavy rains, and wildfires has increased globally in recent decades [3]. Climate change continues to worsen these natural disasters, making them more frequent and more intense [4]. Numerous studies indicate that human-caused carbon dioxide emissions are primarily responsible for climate change and global warming [5]. Nevertheless, to mitigate potential but severe environmental, social, and economic losses, climate change scientists have recommended keeping the increase in global temperature to 1.5°C [6]. To achieve this aim, carbon emissions in the current century must be strictly reduced through a rapidly shrinking "carbon budget" method (i.e., the maximum amount of CO_2 that can be emitted to the atmosphere while limiting global warming to a given degree) [7].

Population growth, urbanization, and built environment expansion have led to global warming. Concrete, the most used artificial material, is needed for infrastructure expansion, with 30 billion tons used annually, surpassing per capita steel and wood consumption [8]. The main ingredient in concrete, cement, has an annual production of over 4 billion tons [9]. Cement manufacturing has raised major environmental concerns because it accounts for 5–8% of global anthropogenic CO_2 emissions. Global process-related CO_2 emissions from the cement sector were 1.57 Gt in 2019 [10], whereas total carbon emissions from cement manufacturing were about 2.9 Gt in 2021 [11].

The cement and concrete industry must decarbonize to meet climate change targets by 2050. Research papers focus on mitigation measures, providing stakeholders with information for net zero roadmaps. This review paper examines the potential of emerging technologies to reduce carbon emissions in the cement sector and suggests research areas for further reduction.

2. Decarbonization strategies

This section reviews major solutions for decarbonizing cement manufacturing, including alternative clinker formulations, fuels, and replacements for clinker in cement production. It also suggests possible research areas to reduce CO_2 emissions in the cement and concrete industries. Below is a diagram of the strategies the cement industry could provide as pathways to net zero by 2050. A top-down strategy was used for the collecting and filtering of publication data, and a mixture of broad keywords was used to extract the preliminary publications. The search was conducted in 2024 and was restricted to English-language peer-reviewed academic papers. "Alternative cement," "alternative clinker," "alternative fuel," "alternative concrete," "carbon emission," "cement carbon sequestration", and "machine learning and AI in cement manufacturing" are some of the broad keywords used in the search process. The methodical approach of the sectors thought to be options for lowering CO_2 emissions in the cement and concrete industries is depicted in **Figure 1**.

2.1 Nanoscale calcium carbonate for gigatonne-scale cement industry decarbonization

Nanoscale calcium carbonate (NCC) are calcium carbonate particles smaller than a micron, ranging from 15–35 to 400 nm [12]. Typically produced as precipitated calcium carbonate (PCC), they present diverse applications across various industries such as adhesives, sealants, food, pharmaceuticals, paints, coatings, paper, and construction materials. Producing higher-value polymorphs of calcium carbonate, like calcite, instead of aragonite or vaterite, can help offset the costs associated with carbon capture, utilization, and storage (CCUS) [13]. Lastly, PCC can also be generated from Ca $(OH)_2$ using ultrasound [14]. Boyjoo et al. [15] offer additional synthesis approaches for both micro and nano-sized calcium carbonate, organizing these first into the CO_2 bubbling method, which sees the most industrial use. Calcite is the

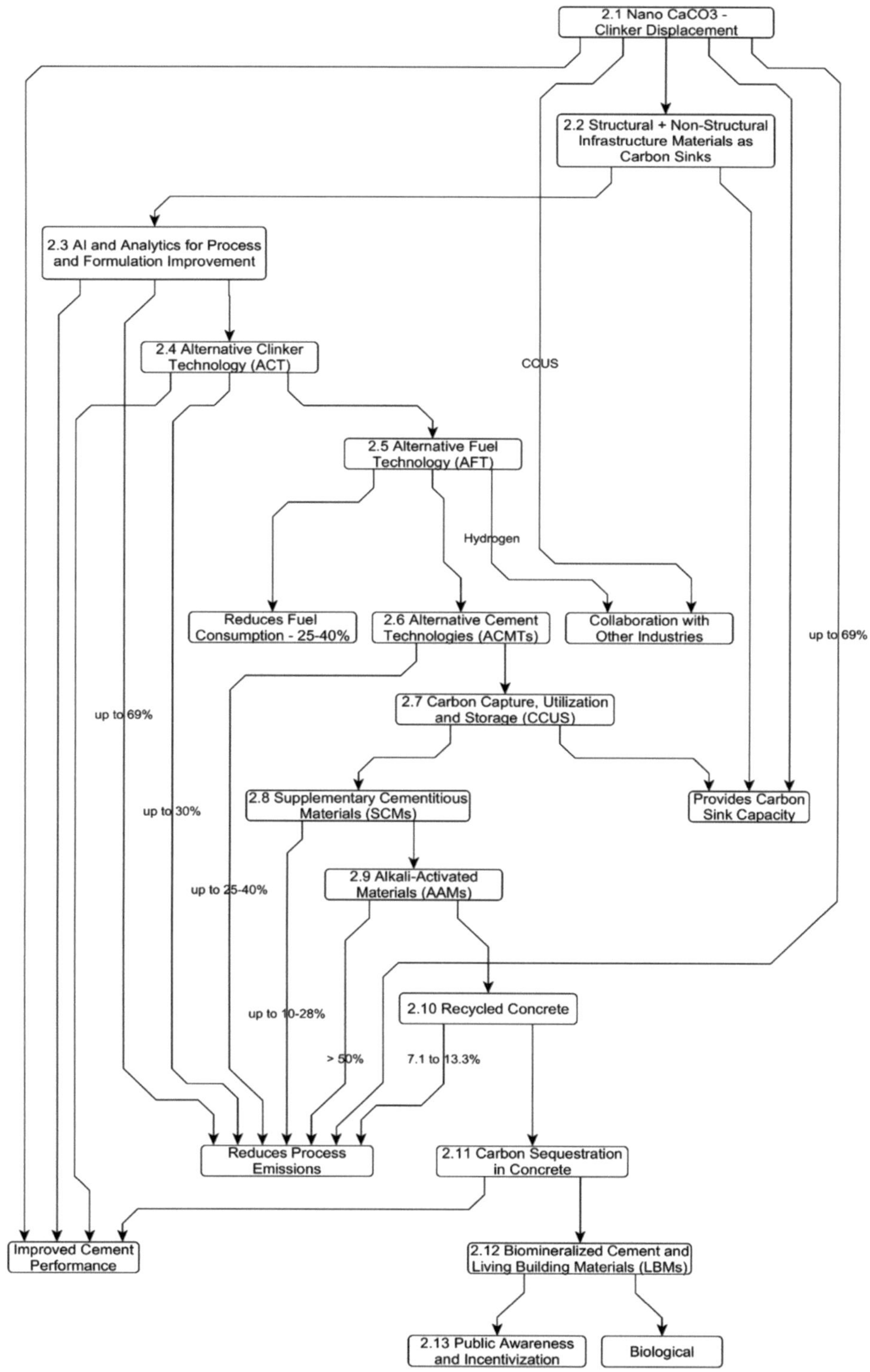

Figure 1.
Shows the major areas considered for possible reduction in CO_2 emission for the cement manufacturing and concrete industry.

primary synthesized polymorph, with NCC more likely than micro and nano-sized particles. Biomimetic approaches involve precipitation and reverse emulsion, separated into spontaneous precipitation and slow carbonation reactions.

NCC in the form of PCC can be made with surfactants with concomitant changes in size, shape, and other properties [12]. These particles can also be sourced from carbon mineralization activities designed to remove CO_2 from point sources or the atmosphere. PCC can also be made with nano-silica using microbes [16] and by using $CaCl_2$ from Solvay process wastewater to remove CO_2 [17]. Liu et al. [16] describe the precautions for managing adverse health outcomes such as silicosis from breathing in nano-silica [18]. The Solvay wastewater carbonation route's scalability depends on wastewater calcium concentration, flow rate, and CO_2 concentration, and has been commercially advanced for carbonating subsurface mineral brines. Displacing clinker is how NCC offers a significant decarbonization potential. According to Batuecas et al. [19], 2% of NCC promises to displace enough clinker to eliminate 69% of global greenhouse gas emissions from the global cement industry. This would be without altering the performance of the resulting cement in terms of its compressive strength or other properties. This potential may be lowered in the case of NCC made from nano-silica and microbes, as described by Liu et al. [16], due to the nutrient input required for microbial growth. Nutrient production in microbial NCC requires energy, increasing carbon footprint. Obtaining nutrients from food waste reduces LCA impact but introduces contamination concerns. Sterilization reduces LCA but requires energy. McDonald et al. [20] describe their carbon-negative manufacturing process for their admixture, PCC-A, as sequestering between 100 and 350 kg CO_2 per tonne of their PCC-A produced. Compressive strength peaks at 10 wt% PCC-A, but workability declines, requiring more water. Still, a significant emissions reduction is the result, which is 27–30%, depending on the alkali used.

Improved performance characteristics suggest value in the use of NCC. These include compressive strength and other performance properties of cement containing NCC, as well as studies demonstrating the viability of NCC in cement production [19]. NCC offers additional routes to decarbonization beyond use in the cement industry. One of those is in improving soil nutrient availability, as demonstrated in fields that grow winter wheat [21]. Another involves its deposition onto nanocellulose to make a flexible material that can make up most paper [22]. NCC is already a bright white compound ideal for the color of paper, and its use would reduce the number of new trees needed. It may even enable the use of more recycled paper materials that could be coated with CC to remove non-uniformities in color, thus providing higher value. Market acceptance and regulatory considerations offer significant hurdles to more widespread adoption. Changing the cement composition carrier has an even higher risk of successful hurdles to adoption. Even with tests showing improvements, representation in databases used by structural engineers allowing for selection by architects on new construction is needed for structural use. Using any new composition can carry significant risks if testing is improperly done—structural failures due to poor quality cement have happened. It is crucial to identify these as project risks that have more to do with the validation and verification of proper manufacture and use of the NCC-containing cement rather than arising from the NCC itself.

2.2 Harnessing structural and non-structural infrastructure materials as carbon sinks

Another promising avenue for mitigating the impact of climate change is using carbon capture strategies, with a particular focus on structural and non-structural infrastructure materials as effective carbon sinks. The potential of these materials to sequester carbon dioxide offers an additional innovative approach, contributing to the reduction of greenhouse gas emissions and the overall fight against climate change. In a transformative materials report, Kreigh [23] advocated for the in-depth examination of certain construction materials for their high carbon-storing potential. These materials include algae-grown bricks, purpose-grown fiber, earthen slabs, non-Portland cement concrete slabs, mycelium structural tubes, and agricultural waste panels. Early-stage lab development of promising materials for building applications, evaluated for durability, fire performance, structural capacity, and thermal conductivity, could be accelerated by policies and legislation.

Structural infrastructure materials can be carbon sinks. Concrete and steel, common in construction, can act as long-term carbon sinks through carbonation, a mineralization process where CO_2 reacts with alkaline components to form stable carbonates. Forced carbonation, increasing CO_2 during early hydration, improves concrete strength, resistance to chloride permeability, freeze-thaw performance, and reduces water absorption [24]. Large structures like dams can be used for carbon sinking without compromising mechanical properties. Carbonation can continue over decades, locking away carbon. Geopolymers offer lower carbon footprints as alternative binders. Bio-based panels integrate seamlessly into construction, storing carbon, offering non-toxicity, supporting local economies, and simplifying production. Clay plaster/panels and algae cement provide fire resistance. Optimizing material formulations is crucial to maximize carbon capture potential.

Non-structural infrastructure materials can also be carbon sinks. The carbon storage capacity of buildings is primarily determined by the quantity and volume of wooden elements incorporated into the structural and non-structural components rather than being significantly impacted by factors such as the building type, wood type, or building size [25]. Beyond traditional construction materials, non-structural infrastructure elements present unique opportunities for carbon capture. Pavements, for instance, can sequester carbon through the absorption of CO_2 during curing. Incorporating recycled materials and exploring alternative pavement designs can enhance the carbon capture potential of roadways. Concrete tetrapods can also be mass-produced for the control of beach erosion, serving as an extensive carbon sink.

Bio-based materials offer versatile solutions for constructing prefabricated panels used in wall and roof enclosure systems, as well as in framing, insulation, and sheathing. These panels seamlessly integrate into current construction practices, store significant carbon, and are non-toxic. Utilizing locally available fiber residues and employing low-tech manufacturing processes enhance their appeal. Specific bio-based panels like clay plaster/panels and algae cement offer fire-resistant properties, enhancing their versatility. Thatch cladding, modernized through mechanized processes, provides durable, affordable, and esthetically pleasing roofing solutions. Carbon-negative products made from natural materials like jute and sisal offer sustainable alternatives for building infrastructures, contributing to environmental and economic development. Research indicates that when treated with alkali solution and compounded with epoxy, these fibers exhibit better mechanical properties suitable for load-bearing composites in construction [26]. Incorporating vegetation into urban landscapes, such as through green roofs and walls, serves as a means of carbon sequestration, offering both environmental and esthetic benefits. Research aimed at identifying optimal plant species, soil conditions, and maintenance practices can enhance the carbon sequestration potential of these features. Alternative materials face durability, economic feasibility, and energy requirements challenges. Life cycle assessments are needed to minimize environmental impact. Infrastructure as carbon sinks may help combat climate change, but ongoing research will be needed to maximize carbon capture and storage potential.

2.3 Use of artificial intelligence and analytics for process and formulation improvement

The emergence of artificial neural networks and machine learning algorithms has enabled developers to incorporate programming that allows decisions on cement formulation alternatives and manufacturing process analysis, similar to how the human brain would process the information. One of the primary benefits of an artificial neural network (ANN) is that it can be trained based on samples (e.g., formulation alternatives, compressive strength evaluations, etc.) and used for classification and measuring correlation. Neural networks can classify information using probability calculations and regression models, with feedback loops for further processing until sufficient certainty is reached. Before performing classification, ANNs require training to develop machine learning capabilities which involve the mapping of various inputs effectively to their respective outputs. The system processes data through the input layer. The classification methodology is based on predictions and decisions, with the number of hidden layers and neurons used to determine the complexity of the task, with careful consideration to avoid over-training and decreasing processing efficiency. Measures of correctness can then be obtained depending on how well the classification has occurred. Weights can then be applied to the confidence level in a specific feature in correctly classifying the object. With each iteration, the ANN learns as weights are modified based on the results of previous classification attempts. This methodology can be applied to analyzing the cement manufacturing process at specific facilities [27] and the contributions of each stage to overall carbon dioxide emissions. As an example, using over 31,000 sensor-based data points and daily CO_2 generation calculations, the cyclone gas outlet temperature and the kiln main drive speed control were identified during one study as the two variables with the most significant correlation with CO_2 generation. Such analytics provide valuable process data on correlations that may not be apparent.

Similarly, various artificial intelligence applications have been applied to the cement manufacturing process for potential improvements. For example, in calculating the apparent degree of calcination, Gang and Hui [28] created a model utilizing a least squares support vector radial basis function (RBF) kerneled machine called the least-squares support-vector machine (LS-SVM). The model's inputs included the furnace's temperature and pressure, the calciner's outlet temperature and pressure, the tertiary air's temperature, and the amount of cement raw that was laid off. To obtain the required level of precalcination of the raw feedstock and low level of carbon monoxide, while stabilizing the precalcination process considering the multi-variable dependencies in the precalciner system, Griparis et al. [29] developed an adaptive, resilient, and fuzzy control design approach To estimate the kiln temperature and oxygen content based on five variables, which are the coal flow to the kiln, coal flow to the precalciner, raw meal flow, rotary speed of the kiln, and negative pressure of the preheater exit, Yang et al. [30] developed a back-propagation neural network (BPNN) and radial basis function neural network (RBFNN).

Machine learning and deep learning have also been used for structural engineering, assessing characteristics such as compressive strength, tensile strength, seismic performance, vibration control, and more. A thorough review of the use of predictive models for concrete properties using machine learning and deep learning was conducted by Moein et al. [31] and depicts how learning models that respond in nonlinear space can be used for analysis when the relationships are nonlinear. The study consolidated test results from various learning methods (deep learning, support vector machine, decision trees, ANN) to evaluate performance metrics such as shear strength, flexural strength, tensile strength, and compressive strength and shared the pros and cons with each methodology used. Artificial intelligence in cement manufacturing is expected to enhance performance and promote sustainable production methods, with further research and development of new variables expected to drive growth.

2.4 Alternative clinker technology (ACT)

The production of one ton of conventional clinker releases about 0.83 tons of carbon [32, 33]. The calcination of limestone and the combustion of fossil fuels are the two primary sources of carbon emissions during clinker production. 60–65% of the carbon emissions connected to the production of clinker are caused by calcination, a high temperature chemical process in which calcium carbonate ($CaCO_3$) is decomposed into calcium oxide (CaO) and carbon dioxide (CO_2). One workable method for achieving decarbonization in the cement sector for medium- to long-term strategic planning is the implementation of alternative clinker technologies (ACT), involving a partial replacement of cement in producing concrete and mortar [34, 35]. In a comprehensive study, Alaloul et al. [36] investigated the addition of oil shale ash (OSA) to cement and geopolymer concretes. More carbonate minerals are often found in oil shale with lower organic content. One oil shale type with major oxide ratios resembling those of OPC clinker is calcareous oil shale [37, 38]. Goncharov and Zhutovsky [38] have demonstrated that calcareous oil shale can substitute more than 76% of the raw materials required to manufacture belite cement clinker. Belite (C_2S) can save carbon emissions by up to 10% since it has a lower CaO content and can be produced at lower calcination temperatures than alite (C_3S).

Oil shale can replace conventional clinker combustion fuel in a rotary kiln, reducing CO_2 emissions [38]. However, current research focuses on technical aspects, requiring further study on embodied carbon analysis and technical performance for accurate scale-up potential. Nehdi et al. [39] offer a thorough review of the literature on alternative clinker technology. Cement with lower carbon emissions has been developed using clinker methods, including magnesium hydroxy-carbonate cement. In 2009, Vlasopoulos and Cheeseman invented this cement, which sequesters CO_2 in hydration products but cannot be low-CO_2 due to its magnesium oxide origin [40, 41]. Therefore, basic research on the energy-efficient industrial process of manufacturing magnesium oxide (MgO) from magnesium silicate rocks is essential [42]. Solidia cement is a different non-hydraulic binder granted a patent in 2016 [43]. A 30% decrease in carbon emissions is possible because of the cement's clinker, which has a composition like OPC clinker but less $CaCO_3$ and a kiln temperature of roughly 1200°C [44]. Furthermore, the curing process of Solidia cement has the potential to absorb 300 kg of CO_2 for every 1000 kg of binder; this process can be expedited at higher temperatures [44, 45]. Solidia cement can only be utilized in precast concrete plants due to its restricted carbon curing technology requirements. Karlsruhe Institute of Technology and SCHWENK Zement KG are the patent holders of the hydraulic binder Celitement [46]. This cement is made from stabilized and synthesized short-lived precursors of CSH, requiring less energy and releasing less carbon dioxide into the atmosphere [45]. Tíecnico-Lisbon and CIMPOR developed and patented X-Clinker, a distinct hydraulic binder [47]. Compared to ordinary OPC, this cement's raw mix has 33% less $CaCO_3$, a lower C/S ratio, and 25% less processing carbon emissions. To create this kind of clinker, the raw combination must be melted at 1550°C during the pyroprocessing phase. The difference in temperature between this and OPC's processing temperature is about 100°C. Consequently, technical modifications to the industrial plants are required to facilitate the development of a 100% liquid phase in the clinker manufacturing process. Refer to Refs. [44, 48, 49] for further technical information on different kinds of clinkers. Literature suggests different ACT phase compositions, but further investigation is needed to understand cement types' engineering performance and contribute to decarbonization through the life cycle assessment of embodied carbon emissions.

2.5 Alternative fuel technology (AFT)

The mean CO_2 emissions from the manufacture of cement range from 563 to 831 kg CO_2/tonne clk worldwide. While energy emissions release 168–476 kg CO_2/tonne clk, calcining limestone releases around 365–560 kg CO_2/tonne clk [45, 50]. According to Chatterjee and Sui [51], the cement industry's global energy consumption and CO_2 emissions in 2016 were predicted to be roughly 11 EJ and 2.2 Gt, respectively. Studies show that switching to zero-emission fuel sources in cement mills might reduce carbon emissions from cement manufacturing by 25–40% [50, 52]. Many alternative fuel technologies, or AFTs, have been implemented to lower carbon emissions associated with the energy required to produce clinker. Three main categories can be used to group these alternative technologies.

Alternative fuels derived from biomass residues from biogenic and non-biogenic processes have been widely used in the cement industry due to their range of resources. Agriculture, manufacturing, packaging, construction and fabrication, food processing and animal husbandry, community and home, transportation, and automotive resources were the categories Chatterjee and Sui [51] divided waste and biomass fuels into. Research studies explore waste and biomass energy compositions, including sewage sludge, which can be combined with Portland cement to create cement or co-combust in a kiln. The latter technology (2016) allows for energy recovery from sewage sludge. Used tires and waste plastic are thought to be two of the readily available energy sources to produce cement; the energy content of plastics is 28–40 MJ/kg. Numerous tire kinds have a chemical composition like that produced in fossil fuel kilns that could be utilized as an alternative fuel for clinker manufacture in cement kilns. Industrial oils and solvents are another important energy source for cement manufacturing, with LHVs ranging from 29 to 36 MJ/kg. Rahman et al. [53] evaluated the heating value and fundamental makeup of waste solvents and heavy fuel oil that can be burned in a cement kiln.

Hydrogen is a desirable clean alternative fuel for future energy-intensive materials research and industrial operations due to its unique properties, which include its complete storage capacity, renewability, zero emissions, adaptability, and quick recovery [54]. The cement and hydrogen fuel industries can collaborate through two methods: burning H_2 for clinker thermal energy and recovering heat from cement manufacturing facilities. Juangsa et al. [55] demonstrated by thermodynamic analysis that, compared to a coal-based plant, the carbon emissions from the carbon-free combustion of H_2 fuel in a cement manufacturing system combined with NH_3 dehydrogenation may be reduced by 44%. El-Emam and Gabriel [56] investigated seven alternative energy mix scenarios where hydrogen fuel partially supplied heat in the kiln. When these hydrogen-based scenarios are used instead of coal-based cement manufacture, they find that carbon emissions can be reduced by 15.0–19.6%. Several notable cement production sites, including Heidelberg Cement in the UK and Cemex in Spain, have conducted hydrogen fuel mix trials. The creation of hydrogen in cement plants is another decarbonization tactic suggested in several studies [57]. Weil et al. [58] proposed a plant design that uses waste gasification to produce hydrogen within a cement production facility. They stated several benefits to the proposed hybrid system, such as reduced manufacturing costs, gasification products instead of primary fuels, and clinker production using fuel ash as a feedstock [58]. Ozturk and Dincer [59] developed an integrated system using waste heat from cement slag and natural gas to produce hydrogen. The cement plant achieved 55% and 22% energy efficiency in its waste heat recovery system, indicating the feasibility of decarbonizing the industry using hydrogen fuel. In their study, Nehdi et al. [39] offer an incredibly comprehensive literature review on this topic.

2.6 Alternative cement technologies (ACMTs)

First developed in the late 1980s, Portland Lime Cement (PLC) technology steadily progressed in the 2000s [60]. Clinker, gypsum, and limestone are interground to create PLC; limestone typically substitutes 10–20% of the OPC clinker [61]. As per the European standard EN 197 (EN BSm, 2011), CEM Type II/B-L PLCs are allowed to have higher replacement levels of up to 35%. Most early studies on PLCs evaluated the material's mechanical and durability properties; hence, the environmental benefits were largely disregarded [62, 63]. As more people realized that blended cement had less influence on the environment, PLCs started to appear in the early 2000s [64, 65]. According to reports, PLC emits 10% less carbon dioxide than OPC [66]. PLCs' clinker factor and carbon emissions can be further reduced using supplemental cementitious materials (SCMs) [67]. Nevertheless, few publicly available research studies contrast how different kinds of PLCs contribute to the cement industry's short-term decarbonization strategies. Nevertheless, the environmental benefits of PCL might be increased even further by replacing a portion of the clinker with calcined clay.

The clinker factor in PLC manufacturing may be further reduced with additional SCMs. Limestone calcined clay cement (LC3) is a potential technique for employing calcined clays as an effective SCM to partially replace clinker in cement production [68]. Consequently, clay is calcined by heating it to 700–850°C. Clays containing kaolinite can be calcined in roller hearth kilns, flash calcination units, and conventional rotary kilns due to their lower calcination temperature than OPC clinker [69]. A blend of limestone and calcined clay might replace a significant amount of the OPC clinker in blended cement. Half of the OPC clinker can usually be replaced with 30% calcined clay, 15% limestone, and 5% gypsum to provide an engineering performance like that of regular OPC. The mechanical, durability, and hydration dynamics of concrete built using LC3 have all been the subject of numerous studies [70]. LC3 has been shown to dramatically reduce carbon emissions from the cement industry [71]. Thus, by replacing 50% of the clinker with calcined clay and limestone, it is possible to reduce carbon emissions by up to 25–40% while maintaining excellent mechanical and durability performances [72, 73]. Clinker-free hydraulic binders are created from various waste types, reducing environmental impact and carbon emissions. Early research substituted essential elements, but studies show it is possible to create hydraulic binders entirely from waste.

2.7 Carbon capture, utilization, and storage (CCUS)

To decarbonize the cement and concrete sectors by 2050, carbon capture, utilization, and storage (CCUS) is an up-and-coming and essential long-term solution. Cement plants [74], concrete production during the curing process [75], and the recycling of cement-based composites that have reached the end of their useful life [76] are all applications of CCUS technology. This section covers the most recent CCUS technologies that apply to cement plants. The IPCC special report on CCUS [77] lists industrial separation, post-combustion, pre-combustion, and oxyfuel combustion as the four primary categories of CO_2 capture methods. The pre-combustion CCUS techniques are mostly coupled with gasification technology to produce hydrogen fuel. However, as noted by Plaza et al. [74], there is a great deal of opportunity for CCUS in the cement industry with direct CO_2 capture [78], oxyfuel combustion [79], and post-combustion [80]. Consequently, the annual CO_2 storage capacity varies from 25,000 to 2 million tons for different methods. For instance, the Holcim Portland Cement Plant in Colorado, USA, intends to operate a trial system capable of capturing two million tons of CO_2 yearly [74]. Cement companies can reduce the linked impact of climate change by 74–91% by switching to oxyfuel combustion technology, claim [81]. It was demonstrated that for each kilogram of clinker generated, using oxyfuel in conjunction with biomass fuels can produce harmful CO_2 emissions of 24 to 169 gCO_2, or equivalent [81]. Since many different technologies are suitable for the cement industry, more research is needed to evaluate the GHG emissions and potential carbon savings for different CCUS systems. For example, Izumi et al. [82] proposed GHG emission calculations for mineral CCUS technology that may be used in LCA inventories. Even though studies have shown that CCUS has a promising future for decarbonizing the cement industry, more thorough life cycle assessment (LCA) research focusing on different technical systems is needed to forecast reductions in carbon emissions more precisely. The US CCUS market could reach USD 4.3 billion–8.5 billion by 2027. However, it is essential to note that this is still an ongoing technology being developed and optimized to be applied in cement manufacturing and even in electrical power generation plants.

2.8 Supplementary cementitious materials (SCMs)

Innovative low- or zero-carbon concrete technologies are being developed to reduce the construction industry's carbon emissions. These technologies focus on enhancing the binder system and waste valorization techniques. SCMs, extracted from various materials, are widely used in the concrete industry as a partial replacement or cement substitute, resulting in significant carbon emissions reduction. Furthermore, using SCMs derived from industrial wastes promotes CE through waste valorization [83]. The main barrier to SCM use is the scarcity of classic SCMs, such as FA, GGBFS, and SF (silica fume). Therefore, it is critical to pinpoint the expanding sources of new SCMs. Juenger et al. [84] assessed the research on innovative SCM resources and their long-term effects on the performance and durability of concrete. The technical aspects of using SCMs in concrete technology have also been the subject of various studies [85]. However, a comprehensive examination of the contributions that various SCM types make to carbon reduction strategies has not yet been carried out.

This section discusses the potential carbon reductions of SCMs made from different resources. Industrial waste is the source of the SCMs most frequently used in concrete production. SF, GGBFS, and FA are the additives that are most frequently utilized in cement and concrete production worldwide. The effects of using these SCMs in cement and concrete production have been the subject of numerous investigations [85, 86]. SCMs improve the engineering properties of concrete while also having a positive impact on the environment [87]. For instance, Tushar et al. [88] discovered that substituting 50% of the world's cement output with FA and GGBFS can save 209 billion dollars and reduce 2745 million tons of CO_2 emissions. Using various traditional SCMs can lower greenhouse gas emissions by 10–28% and the global warming potential (GWP) by 20–38%, according to Hossain et al. [89]. Comparable results were also found in similar experiments [90]. Conventional supply chain management (SCM)'s integration into net-zero carbon goals remains a contentious issue due to constraints like the absence of popular industrial leftover-derived solutions. Miller [9] asserts that the distance and mode of transportation of SCMs may overshadow their contribution to greenhouse gas emissions, making the strategic deployment of SCMs particularly crucial. Arrigoni et al. [91] indicated that SCMs with an unfettered market can minimize GHG emissions in most circumstances, highlighting the significance of SCMs' market limits in LCA research. Inefficient use of restricted SCM alternatives, particularly when transported over long distances, can potentially exceed the GHG emissions of standard OPC concrete. Additionally, they discovered, in keeping with Miller's [9] findings, that lowering SCMs does not always translate into a decrease in concrete's greenhouse gas emissions.

2.9 Alkali-activated materials (AAMs)

Geopolymers and other alkali-activated materials (AAMs) are recognized as environmentally friendly building materials that drastically reduce carbon emissions. Though they might only partially replace OPC concrete globally, AAMs have a strong chance of serving as an alternative concrete technology in regions with locally available raw materials. An aqueous solution of an alkali hydroxide, silicate, carbonate, or sulfate activator is required for AAMs to function. These AAMs are typically divided into two categories: one-part (also called "just add water") and two-part (sometimes called "traditional") AAMs. A dry combination comprising a solid aluminosilicate precursor and a solid alkali-activator is added to water to create the binder matrix in one-part mixes [92]. A range of aluminosilicate binders, including FA, GGBFS, metakaolin, and calcined clay, can be the precursor in AAMs.

Although numerous studies in the literature [93, 94] assessed the technical aspects of developing AAMs, this section concentrates on the potential of AAMs in lowering carbon emissions. In contrast to the other alternative technologies investigated in this study, life cycle assessments (LCAs) in the public domain have been used to investigate the prospect of lowering AAM carbon emissions in detail [95]. AAMs can save CO_2 emissions by more than 50% compared to OPC concrete, making them a good option for short-term decarbonization initiatives. The findings of this study, however, emphasize the critical role that activators play in determining the amount of carbon that AAMs can preserve. According to Robayo-Salazar et al. [96] and Fernando et al. [97], alkali activators such as sodium silicate and sodium hydroxide are responsible for most carbon emissions in AAMs.

Therefore, while creating low-carbon AAMs, it is crucial to carefully consider using chemical admixtures, especially when the target strength is high. A further important factor in producing AAMs with significantly lower carbon footprints is the availability of local raw materials for use as precursors. The transportation of raw materials and the use of low-reactivity precursors may increase the carbon footprint of AAMs. One potential replacement for carbon-intensive activators in one-part AAMs is using suitable industrial wastes. Adesanya et al. [98] used desulfurization dust instead of store-bought sodium hydroxide activators. They offered evidence that using these types of industrial wastes can create AAMs that are cleaner and have a lower carbon footprint. Similar outcomes were also seen when one-part AAMs were activated with RGP, RHA, and lime kiln dust [99]. More research is required to determine how newly developed AAMs that substitute various types of waste for chemical activators would affect the environment.

2.10 Recycled concrete

Reusing leftover concrete in environmentally friendly buildings is a common practice. Recent studies explore the use of leftover cement paste to create recycled cement. Recycling construction and demolition waste (CDW) is crucial for sustainability. Recycled aggregate concrete (RAC) is made by substituting RA for NA, with progress in mechanical and durability characteristics reviewed [100, 101]. Furthermore, fundamental research has been conducted to assess the extent to which RAC technology lessens the environmental impact of the concrete industry [90, 102]. RAC technology can decrease concrete production's carbon footprint, but factors like transportation distance to landfills and waste disposal facilities are crucial for determining success [103]. RA instead of NA may not significantly reduce carbon emissions but eliminating long transit of CWD or NA can.

Carbonated RAC technology can reduce emissions and improve RA quality through carbon mineralization. According to Tan et al. [104], RA carbon-conditioning is a practical method that provides a feasible usage for CRAC. Xiao et al. [105] found that concrete made with carbonated RA had 7.1–13.3% lower carbon emissions than concrete made with NA or non-carbonated RA, respectively, due to the possibility of carbon absorption in carbonated RA. An interesting option to increase carbon savings is to combine RAC and CRAC with another waste valorization technique, such as employing by-products as SCMs or precursors for AAMs [106, 107]. Research shows recycled concrete can decarbonize the concrete industry, with innovative applications like reusing it as filler in tetrapods, promoting sustainable waste utilization.

2.11 Carbon sequestration in concrete

Carbon sequestration in concrete curing can reduce carbon footprint by capturing CO_2 and forming stable calcium carbonate. Early applications show promising results [108]. Several innovative CCUS technologies have been proposed for the concrete industry, including carbonation of recycled aggregate concrete (CRAC) and CO_2 mineralization in synthetic cementitious materials (SCMs) [109]. While contemporary CCUS technology holds potential for carbon sequestration, challenges remain regarding the environmental impact and effectiveness of these processes. For example, while adding ground granulated blast furnace slag (GGBFS) to cement can result in significant carbon absorption without compromising compressive strength, concerns exist regarding water usage and achieving full carbonation throughout the concrete's depth [110]. Some studies have even questioned the net CO_2 benefit of CCU concrete, emphasizing the need for further research to optimize electricity usage and enhance compressive strength during carbon curing [75, 111]. Thorough experimental and life cycle assessment (LCA) studies are essential to accurately evaluate the carbon-saving potential of various CCUS technologies in the concrete sector. Biochar, derived from biomass pyrolysis, has the potential to sequester up to 2.6 tons of CO_2 per ton [112].

Studies indicate that adding biochar to cement at low concentrations enhances hydration and compressive strength [113]. While its impact on concrete properties has been extensively studied, research on its carbon-sequestering potential is limited. Nevertheless, biochar incorporation can support circular economy initiatives and reduce concrete's carbon footprint [114]. Factors influencing its carbon capture effectiveness include activation methods and manufacturing characteristics [115]. Recent studies suggest that concrete with biochar and supplementary cementitious materials can achieve carbon-negative status [10]. However, further research is needed to explore composite properties and optimize formulations [116]. Overall, biochar shows promise in reducing concrete's carbon footprint, but comprehensive life cycle assessments are necessary for a thorough understanding of its efficacy.

2.12 Biomineralized cement and living building materials (LBMs)

Microbial biomineralization is the basis for developing bio-cement technology, which aims to improve the sustainability of newly produced building materials. The utilization of calcium carbonate biomineralization processes in construction materials was reviewed by Beatty et al. [117]. Among biomineralization methods, engineered living building materials (LBM) provide a novel technology that adds various functionalities to building materials using biology [118]. Sand-gelatin structural scaffolds are biomineralized by photosynthetic bacteria that can produce microbially induced calcium carbonate precipitation (MICP), such as *Synechococcus* sp. [119] and *Escherichia coli*, to create LBMs [120]. As a result, LBMs can acquire strength without the use of cement. This method has garnered attention because CO_2 sequestration during cell growth has such a promising potential. During photosynthesis, sequestered CO_2 is liberated and reacts with water to produce biomineral calcium carbonate. According to Heveran et al. [119], LBM is a technology that gives building materials multifunctionality sensing, responsiveness, and regeneration through biology. However, since LBM is still in its infancy, more study is needed to determine its scalability and other possible uses, like carbon sequestration [121]. Additional uses for biomineralization include better self-healing cement-based composites and bacterial treatment of waste-based cementitious materials. According to Sharma et al. [122], adding up to 40% sandstone powder to mortars and bacterial treatment can enhance mechanical and durability while reducing carbon emissions. The ability of LBMs to self-heal was investigated by Delesky et al. [123] using biomineralizing *Synechococcus* sp. PCC 7002. Biomineralization, a new technology, can lower cement consumption in the building sector by promoting calcium biological carbonate precipitation for self-healing, repair, and rehabilitation of cement-based composites or creating cement-free LBMs.

2.13 Public awareness and incentivization

Equal to the level of effort taken to improve cement manufacturing processes and formulations is the need for public awareness to enforce these changes through the procurement of more eco-friendly products. Colorado and California have recently incorporated legislation for environmental product declarations to include the amount of CO_2 produced while manufacturing construction materials such as cement and asphalt. Incentivization efforts such as product rating systems can help consumers share in the efforts for increased sustainable production. The Concrete Sustainability Council is one such organization offering a certification system designed to provide concrete, cement, and aggregate companies insight into the levels at which source ingredient companies operate in an environmentally, socially, and economically responsible way [124]. Rating systems like LEED and Greenroads promote environmentally friendly construction materials, while EnergyStar certifications assess cement efficiency. Promoting sustainable manufacturing requires differentiators and public awareness.

3. Discussion

In Section 2, 13 different approaches were described together with ranges of CO_2 emissions reduction percentages; these are listed in **Table 1**. **Table 1** shows the ongoing exploration of CO_2 emission reduction solutions in the cement and concrete sector. However, most technologies are in the early stages, and it's unclear how much CO_2 will be reduced. CCUS technology promises a 91% reduction. None of the CCUS systems are currently operational or under study with promising funding, while other solutions aim to reduce CO_2 emissions by over 50%, but this does not necessarily mean a significant reduction. A 50% reduction for the entire cement sector may be less than the 69% estimated for the use of NCC, but 50% of 6 Gt yields a much larger amount of carbon emissions saved than 50% of a million tonnes. Reporting values that share a common point of reference, such as the annual emissions of a cement kiln per tonne of

Section	Review findings	CO_2 emissions reduction percentage
2.1 Nano $CaCO_3$—Clinker displacement	Ref. [15]. The CO_2 bubbling method is widely used in industry for creating microsizes and nanosizes of calcium carbonate.	Up to 69%
2.2 Structural and non-structural infrastructure materials as carbon sinks	Ref. [24]. Forced carbonation, is a process that increases CO_2 during early hydration, enhances concrete strength, resistance to chloride permeability, freeze-thaw performance, and reduces water absorption.	Not quantified (new technology)
2.3 AI and analytics for process and formulation improvement	Ref. [31]. The study examined the utilization of machine learning and deep learning in predicting concrete properties.	Not quantified (new technology)
2.4 Alternative clinker technology (ACT)	Ref. [44]. Solidia cement's curing process can absorb 300 kg of CO_2 for 1000 kg of binder and can be expedited at higher temperatures.	Up to 30%
2.5 Alternative fuel technology (AFT)	Ref. [50]. Studies indicate that transitioning to zero-emission fuel sources in cement mills could potentially decrease carbon emissions by 25–40%.	Up to 25–40%
2.6 Alternative cement technologies (ACMTs)	Ref. [67]. PLCs' clinker factor and carbon emissions can be further reduced using supplemental cementitious materials (SCMs)	Up to 25–40%
2.7 Carbon capture, utilization, and storage (CCUS)	Ref. [81]. Cement companies can reduce the linked impact of climate change by 74–91% by switching to oxyfuel combustion technology, claim.	The current promise is up to 91%
2.8 Supplementary cementitious materials (SCMs)	Ref. [89]. Using various traditional SCMs can lower greenhouse gas emissions by 10–28% and the global warming potential (GWP) by 20–38%.	Up to 10–38%

Section	Review findings	CO_2 emissions reduction percentage
2.9 Alkali-activated materials (AAMs)	Refs. [95–97]. AAMs can save CO_2 emissions by more than 50% compared to OPC concrete, making them a good option for short-term decarbonization initiatives.	>50%
2.10 Recycled concrete	Ref. [105] concrete made with carbonated RA had 7.1–13.3% lower carbon emissions than concrete made with NA or non-carbonated RA, respectively, due to the possibility of carbon absorption in carbonated RA	Between 7.1% and 13.3%
2.11 Carbon sequestration in concrete	Ref. [112]. Biochar, derived from biomass pyrolysis, has the potential to sequester up to 2.6 tons of CO_2 per ton	Up to 2.6 tons of CO_2 per ton of biochar
2.12 Biomineralized cement and living building materials (LBMs)	Refs. [119, 121]. LBM technology offers multifunctional building materials sensing, responsiveness, and regeneration through biology, but further study is needed to determine scalability and potential uses like carbon sequestration.	Not quantified (new research area)
2.13 Public awareness and incentivization	Ref. [124] companies in the concrete, cement, and aggregate industries insights into their environmental, social, and economic responsibility.	N/A

Table 1.
Shows the summary percentage of CO_2 emission reduction in the 13 areas presented.

clinker made, can be very helpful. The second item refers to the values of the emissions reduction percentages above. As a very rough approximation, one can add the percentages to determine a range of emissions reductions achievable from 57% to 161%.

In recent times, the cement and concrete industry has heavily focused on Alternative Fuel Technology (AFT), Supplementary Cementitious Materials (SCMs), and Recycled Concrete as solutions for the reduction of emission of CO_2. However as clearly shown in **Table 1** these options of solution alone will not help deal with the CO_2 emission problem. To reach the goal of NetZero by 2050 the industry must heavily invest in more alternative solutions and look at which ones are viable. CCUS is the most promising since it captures most of the CO_2 emission at the manufacturing stage. The other alternative for concrete can be used as CO_2 sequestration to remove CO_2 from the atmosphere.

4. Conclusion

This methodical, in-depth, and critical review looks at the emerging alternative technologies to decarbonize the cement and concrete sectors. Most of the recommended technologies are still in their early stages of development. Therefore, further experimental and LCA research is still needed to expand their applicability. Since most research on low- and zero-carbon technologies has focused on technological advancements, a concerted effort is still required to ascertain the specific contribution of each technology to carbon neutrality in the cement and concrete industries.

The cement and concrete industries can achieve carbon neutrality by integrating a circular economy (CCUS) with low-carbon technology, utilizing waste appreciation, and upcycling unused concrete to minimize their carbon footprint. At this point, it seems impossible to imagine achieving net-zero cement and concrete by 2050. Given the grave threat posed by climate change, there is great hope that the cement and concrete sector will be able to fulfill its commitments and decarbonize. Strict regulatory rules, innovative design, production advancement, innovative technologies, new scientific breakthroughs, and unwavering dedication from all stakeholders can accomplish this.

Acknowledgements

The authors would like to thank the researchers at the Systems Engineering Department of Colorado State University for their support.

Conflict of interest

The authors have no conflict of interest in the publication of this work.

Author details

Kwaku Boakye[1,2]*, Dahl Winters[1], Olurotimi Oguntola[1], Kevin Fenton[1] and Steve Simske[1]

1 Department of Systems Engineering, Colorado State University, Fort Collins, CO, USA

2 Heidelberg Materials Inc., Irving, TX, USA

*Address all correspondence to: kwaku.boakye@colostate.edu

References

[1] Mahlia T. Emissions from electricity generation in Malaysia. Renewable Energy. 2002;**27**:293-300

[2] Zhang XP, Tan YK, Tan QL, Yuan JH. Decomposition of aggregate CO_2 emissions within a joint production framework. Energy Economics. 2012;**34**:1088-1097

[3] Frederikse T, Buchanan MK, Lambert E, Kopp RE, Oppenheimer M, Rasmussen D, et al. Antarctic ice sheet and emission scenario controls on 21st-century extreme sea-level changes. Nature Communications. 2020;**11**:1-11

[4] Coffetti D, Crotti E, Gazzaniga G, Carrara M, Pastore T, Coppola L. Pathways towards sustainable concrete. Cement and Concrete Research. 2022;**154**:106718

[5] Caldeira K, Wickett ME. Anthropogenic carbon and ocean pH. Nature. 2003;**425**:365

[6] Projected Climate Change. Global Warming of 1.5°C. Geneva, Switzerland: World Meteorological Organization; 2018

[7] Masson-Delmotte V, Zhai P, Pörtner HO, Roberts D, Skea J, Shukla PR, et al. IPCC: Summary for policymakers. In: Global Warming of 1.5°C. An IPCC Special Report on the Impacts of Global Warming of 1.5°C above Pre-industrial Levels and Related Global Greenhouse Gas Emission Pathways, in the Context of Strengthening the Global Response to the Threat of Climate Change, Sustainable Development, and Efforts to Eradicate Poverty. Geneva, Switzerland: World Meteorological Organization; 2018. p. 32

[8] Monteiro PJ, Miller SA, Horvath A. Towards sustainable concrete. Nature Materials. 2017;**16**:698-699

[9] Miller SA. Supplementary cementitious materials to mitigate greenhouse gas emissions from concrete: Can there be too much of a good thing? Journal of Cleaner Production. 2018;**178**:587-598

[10] Chen L, Zhang Y, Wang L, Ruan S, Chen J, Li H, et al. Biochar-augmented carbon-negative concrete. Chemical Engineering Journal. 2022;**431**:133946

[11] Chaudhury R, Sharma U, Thapliyal P, Singh L. Low-CO_2 emission strategies to achieve net zero target in cement sector. Journal of Cleaner Production. 2023;**417**:137466

[12] El-Sheikh SM, El-Sherbiny S, Barhoum A, Deng Y. Effects of cationic surfactant during the precipitation of calcium carbonate nano-particles on their size, morphology, and other characteristics. Colloids and Surfaces A: Physicochemical and Engineering Aspects. 2013;**422**:44-49. DOI: 10.1016/j.colsurfa.2013.01.020

[13] Chang R, Choi D, Kim MH, Park Y. Tuning crystal polymorphisms and structural investigation of precipitated calcium carbonates for CO_2 mineralization. ACS Sustainable Chemistry and Engineering. 2016;**5**(2):1659-1667. DOI: 10.1021/acssuschemeng.6b02411

[14] Aldea S, Snåre M, Eränen K, Grenman H, Rautio AR, Kordás K, et al. Crystallization of nano-calcium carbonate: The influence of process parameters. Chemie Ingenieur Technik. 2016;**88**(11):1609-1616. DOI: 10.1002/cite.201600028

[15] Boyjoo Y, Pareek VK, Liu J. Synthesis of micro and nano-sized calcium

carbonate particles and their applications. Journal of Materials Chemistry A. 2014;**2**(35):14270-14288. DOI: 10.1039/C4TA02070G

[16] Liu X, Chen L, Liu A, Wang X. Effect of nano-$CaCO_3$ on properties of cement paste. Energy Procedia. 2012;**16**:991-996. DOI: 10.1016/j.egypro.2012.01.158

[17] Farrag M et al. On the role of floodplain storage and hydrodynamic interactions in flood risk estimation. Hydrological Sciences Journal. 2022;**67**(4):508-534. DOI: 10.1080/02626667.2022.2030058

[18] D'Amora M, Liendo F, Deorsola FA, Bensaid S, Giordani S. Toxicological profile of calcium carbonate nanoparticles for industrial applications. Colloids and Surfaces, B: Biointerfaces. 2020;**190**:110947. DOI: 10.1016/j.colsurfb.2020.110947

[19] Batuecas E, Liendo F, Tommasi T, Bensaid S, Deorsola FA, Fino D. Recycling CO_2 from flue gas for $CaCO_3$ nanoparticles production as cement filler: A Life Cycle Assessment. Journal of CO_2 Utilization. 2021;**45**:101446. DOI: 10.1016/j.jcou.2021.101446

[20] McDonald L, Glasser F, Imbabi M. A new, carbon-negative precipitated calcium carbonate admixture (PCC-A) for low carbon Portland cements. Materials. 2019;**12**(4):554. DOI: 10.3390/ma12040554

[21] Yu G, Dong C, Chen S, Li Y, Shi Y. Effect of nano carbon and nano calcium carbonate application on soil nutrient dynamics in winter wheat (*Triticum aestivum* L.). Communications in Soil Science and Plant Analysis. 2023;**54**:1-13. DOI: 10.1080/00103624.2023.2241502

[22] Han JS, Jung SY, Kang DS, Seo YB. Development of flexible calcium carbonate for papermaking filler. ACS Sustainable Chemistry and Engineering. 2020;**8**(24):8994-9001. DOI: 10.1021/acssuschemeng.0c01593

[23] Kriegh J, Magwood C, Srubar W, Lewis M, Simonen K. Transformative Carbon-Storing Materials: Accelerating an Ecosystem Report. Creative Commons Attribution 4.0 International License; 2021. Available from: http://hdl.handle.net/1773/48126, https://carbonleadershipforum.org/transformative-materials/

[24] Rostami V, Shao Y, Boyd AJ. Carbonation curing versus steam curing for precast concrete production. Journal of Materials in Civil Engineering. 2012;**24**(9):1221-1229. DOI: 10.1061/(ASCE)MT.1943-5533.0000462

[25] Amiri A, Ottelin J, Sorvari J, Junnila S. Cities as carbon sinks-classification of wooden buildings. Environmental Research Letters. 2020;**15**(9):94076. DOI: 10.1088/1748-9326/aba134

[26] Ekundayo G, Adejuyigbe S. Reviewing the development of natural fiber polymer composite: A case study of sisal and jute. American Journal of Mechanical and Material Engineering. 2019;**3**(1):1-10

[27] Boakye K, Fenton K, Simske S. Machine learning algorithm to predict CO_2 using a cement manufacturing historic production variables dataset: A case study at union bridge plant, Heilderberg Materials, Maryland. Journal Manufacturing Material Process. 2023;**199**(7):1-10

[28] Gang YZ, Hui L. Soft sensor for apparent degree of calcination in NSP cement production line. In: Presented at the 2nd International Conference on Computer and Automation Engineering (ICCAE), 2010 The 2nd International Conference on. Volume: 1; February

26-28, 2010; Singapore. IEEE Xplore; 2010. DOI: 10.1109/ICCAE.2010.5451912

[29] Griparis MK, Koumboulis FN, Machos NS, Marinos I. Precalcination in cement plants (system description and control trends). IFAC Proceedings. 2000;**33**:273-278

[30] Yang B, Lu H, Chen L. BPNN and RBFNN-based modeling analysis and comparison for cement calcination process. In: Presented at the 2010 Third International Workshop on Advanced Computational Intelligence (IWACI 2010); August 25-27, 2010; Suzhou, China. IEEE; 2010. IEEE Catalog Number: ISBN: CFP1040J-PRT 978-1-4244-6334-3

[31] Moein M, Saradar A, Rahmati K, Mousavinejad G, et al. Predictive models for concrete properties using machine learning and deep learning approaches: A review. Journal of Building Engineering. 2023;**63**:105444. DOI: 10.1016/j.jobe.2022.105444

[32] Prakasan S, Palaniappan S, Gettu R. Study of energy use and CO_2 emissions in the manufacturing of clinker and cement. Journal of Instant Engineering: Series A. 2020;**101**:221-232

[33] International Energy Agency (IEA). Low-Carbon Transition in the Cement Industry. OECD/International Energy Agency; 2018. Available from: www.iea.org. DOI: 10.1787/9789264300248-en. ISSN: 22182837

[34] Khedaywi T, Yeginobali A, Smadi M, Cabrera J. Pozzolanic activity of Jordanian oil shale ash. Cement and Concrete Research. 1990;**20**:843-852

[35] Smadi MM, Haddad RH. The use of oil shale ash in Portland cement concrete. Cement Concrete Composition. 2003;**25**:43-50

[36] Alaloul WS, Al Salaheen M, Malkawi AB, Alzubi K, Al-Sabaeei AM, Musarat MA. Utilizing of oil shale ash as a construction material: A systematic review. Construction and Building Materials. 2021;**299**:123844

[37] Minster T. Oil Shale Occurrences in Israel: Geological Updated Information. Samuel Neaman Institute for Advanced Studies in Science and Technology, Technion-Israel Institute of Technology; 2013. Available from: https://www.neaman.org.il/Files/6-392.pdf

[38] Goncharov A, Zhutovsky S. Eco-friendly belite cement from crude calcareous oil shale with low calorific value. Cement and Concrete Research. 2022;**159**:106874

[39] Nehdi M, Duquette J, El Damatty A. Performance of rice husk ash produced using a new technology as a mineral admixture in concrete. Cement and Concrete Research. 2003;**33**:1203-1210

[40] Gartner E, Sui T. Alternative cement clinkers. Cement and Concrete Research. 2018;**114**:27-39

[41] Gartner EM, Macphee DE. A physico-chemical basis for novel cementitious binders. Cement and Concrete Research. 2011;**41**:736-749

[42] Gartner E, Hirao H. A review of alternative approaches to the reduction of CO_2 emissions associated with the manufacture of the binder phase in concrete. Cement and Concrete Research. 2015;**78**:126-142

[43] Quinn S, Sahu S. Compositions and methods for controling setting of carbonatable calcium silicate cements containing hydrating materials. Google Patents; 2019

[44] Meyer V, de Cristofaro N, Bryant J, Sahu S. Solidia cement an example of

carbon capture and utilization. Key Engineering Materials (Trans Tech Publ). 2018;**761**:197-203. DOI: 10.4028/www.scientific.net/KEM.761.197

[45] Antunes M, Santos RL, Pereira J, Rocha P, Horta RB, Colaço R. Alternative clinker technologies for reducing carbon emissions in cement industry: A critical review. Materials. 2021;**15**:209

[46] Beuchle G, Stemmermann P, Schweike U, Garbev K. Single-phase hydraulic binder, methods for the production thereof and building material produced therewith. Google Patents; 2013

[47] Horta RSB, Colaco RAC, Lopes JNAC, Dos Santos RL, Pereira JC, Lebreiro SMM. Amorphous low-calcium content silicate hydraulic binders and methods for their manufacturing. Google Patents; 2019

[48] Paradiso P, Santos R, Horta R, Lopes J, Ferreira P, Colaço R. Formation of nanocrystalline tobermorite in calcium silicate binders with low C/S ratio. Acta Materialia. 2018;**152**:7-15

[49] Santos RL, Horta RB, Pereira J, Nunes TG, Rocha P, Colaço R. Alkali activation of a novel calcium-silicate hydraulic binder with CaO/SiO2= 1.1. Journal of the American Ceramic Society. 2018;**101**:4158-4170

[50] Nhuchhen DR, Sit SP, Layzell DB. Alternative fuels co-fired with natural gas in the pre-calciner of a cement plant: Energy and material flows. Fuel. 2021;**295**:120544

[51] Summerbell DL, Barlow CY, Cullen JM. Potential reduction of carbon emissions by performance improvement: A cement industry case study. Journal of Cleaner Production. 2016;**135**:1327-1339

[52] Chatterjee A, Sui T. Alternative fuels–effects on clinker process and properties. Cement and Concrete Research. 2019;**123**:105777

[53] Rahman A, Rasul M, Khan MMK, Sharma S. Recent development on the uses of alternative fuels in cement manufacturing process. Fuel. 2015;**145**:84-99

[54] Qazi UY. Future of hydrogen as an alternative fuel for next-generation industrial applications; Challenges and expected opportunities. Energies. 2022;**15**:4741

[55] Juangsa FB, Cezeliano AS, Darmanto PS, Aziz M. Thermodynamic analysis of hydrogen utilization as alternative fuel in cement production. South African Journal of Chemical Engineering. 2022;**42**:23-31

[56] El-Emam RS, Gabriel KS. Synergizing hydrogen and cement industries for Canada's climate plan–case study. Energy Sources, Part A Recovery, Utilization Environmental Efficiency. 2021;**43**:3151-3165

[57] Nhuchhen DR, Sit SP, Layzell DB. Decarbonization of cement production in a hydrogen economy. Applied Energy. 2022;**317**:119180

[58] Weil S, Hamel S, Krumm W. Hydrogen energy from coupled waste gasification and cement production—A thermochemical concept study. International Journal of Hydrogen Energy. 2006;**31**:1674-1689

[59] Ozturk M, Dincer I. Utilization of waste heat from cement plant to generate hydrogen and blend it with natural gas. International Journal of Hydrogen Energy. 2022;**47**(48):20695-20704

[60] Tsivilis S, Chaniotakis E, Badogiannis E, Pahoulas G, Ilias A. A study on the parameters affecting the properties of Portland limestone cements. Cement and Concrete Composites. 1999;**21**:107-116

[61] Tsivilis S, Tsantilas J, Kakali G, Chaniotakis E, Sakellariou A. The permeability of Portland limestone cement concrete. Cement and Concrete Research. 2003;**33**:1465-1471

[62] Nadelman E, Kurtis K. Durability of Portland-limestone cement-based materials to physical salt attack. Cement and Concrete Research. 2019;**125**:105859

[63] Voglis N, Kakali G, Chaniotakis E, Tsivilis S, Portland-limestone cements. Their properties and hydration compared to those of other composite cements. Cement Concrete Composition. 2005;**27**:191-196

[64] Huntzinger DN, Eatmon TD. A lifecycle assessment of Portland cement manufacturing: Comparing the traditional process with alternative technologies. Journal of Cleaner Production. 2009;**17**:668-675

[65] Barrett TJ, Sun H, Nantung T, Weiss WJ. Performance of Portland limestone cements. Transportation Research Record. 2014;**2441**:112-120

[66] Bushi L, Meil J. An Environmental Life Cycle Assessment of Portland-Limestone and Ordinary Portland Cements in Concrete. Cement Association of Canada; 2014. p. 10. Available from: https://www.athenasmi.org/wp-content/uploads/2014/01/CAC_PLCvsOPC_Final_Technical_Brief.pdf

[67] Gupta S, Mohapatra B, Bansal M. A review on development of Portland limestone cement: A step towards low carbon economy for Indian cement industry. Current Research in Green and Sustainable Chemistry. 2020;**3**:100019

[68] Scrivener KL. Options for the future of cement. Indian Concrete Journal. 2014;**88**:11-21

[69] Scrivener K, Martirena F, Bishnoi S, Maity S. Calcined clay limestone cements (LC3). Cement and Concrete Research. 2018;**114**:49-56

[70] Zunino F, Scrivener K. Microstructural developments of limestone calcined clay cement (LC3) pastes after long-term (3 years) hydration. Cement and Concrete Research. 2022;**153**:106693

[71] Díaz YC, Berriel SS, Heierli U, Favier AR, Machado IRS, Scrivener KL, et al. Limestone calcined clay cement as a low-carbon solution to meet expanding cement demand in emerging economies. Development Engineering. 2017;**2**:82-91

[72] Zunino F, Martirena F, Scrivener K. Limestone calcined clay cements (LC3). ACI Materials Journal. 2021;**118**(3):49. Document: 20-007. ISSN 0889-325X. DOI: 10.14359/51730422

[73] Berriel SS, Favier A, Domínguez ER, Machado IS, Heierli U, Scrivener K, et al. Assessing the environmental and economic potential of limestone calcined clay cement in Cuba. Journal of Cleaner Production. 2016;**124**:361-369

[74] Plaza MG, Martínez S, Rubiera F. CO_2 capture, use, and storage in the cement industry: State of the art and expectations. Energies. 2020;**13**:5692

[75] Zhang D, Ghouleh Z, Shao Y. Review on carbonation curing of cement-based materials. Journal of CO_2 Utilization. 2017;**21**:119-131

[76] Zajac M, Skocek J, Skibsted J, Haha MB. CO_2 mineralization of demolished concrete wastes into a supplementary cementitious material—A new CCU approach for the cement industry. RILEM Technical Letters. 2021;**6**:53-60

[77] Metz B, Davidson O, De Coninck H, Loos M, Meyer L. IPCC Special Report

on Carbon Dioxide Capture and Storage. Cambridge: Cambridge University Press; 2005

[78] Hills TP, Sceats M, Rennie D, Fennell P. LEILAC: Low-cost CO_2 capture for the cement and lime industries. Energy Proceedings. 2017;**114**:6166-6170

[79] Faria DG, Carvalho MM, Neto MR, de Paula EC, Cardoso M, Vakkilainen EK. Integrating oxyfuel combustion and power-to-gas in the cement industry: A process modeling and simulation study. International Journal of Greenhouse Gas Control. 2022;**114**:103602

[80] Majchrzak-Kucęba I et al. Experimental investigation into CO_2 capture from the cement plant by VPSA technology using zeolite 13X and activated carbon. Journal of CO_2 Utilization. 2022;**61**:102027

[81] Cavalett O, Watanabe MD, Fleiger K, Hoenig V, Cherubini F. LCA and negative emission potential of retrofitted cement plants under oxyfuel conditions at high biogenic fuel shares. Scientific Reports. 2022;**12**:1-14

[82] Izumi Y, Iizuka A, Ho H-J. Calculation of greenhouse gas emissions for a carbon recycling system using mineral carbon capture and utilization technology in the cement industry. Journal of Cleaner Production. 2021;**312**:127618

[83] Lothenbach B, Scrivener K, Hooton R. Supplementary cementitious materials. Cement and Concrete Research. 2011;**41**:1244-1256

[84] Juenger MC, Snellings R, Bernal SA. Supplementary cementitious materials: New sources, characterization, and performance insights. Cement and Concrete Research. 2019;**122**:257-273

[85] Juenger MC, Siddique R. Recent advances in understanding the role of supplementary cementitious materials in concrete. Cement and Concrete Research. 2015;**78**:71-80

[86] Snellings R, Mertens G, Elsen J. Supplementary cementitious materials. Reviews in Mineralogy and Geochemistry. 2012;**74**:211-278

[87] Yang K-H, Jung Y-B, Cho M-S, Tae S-H. Effect of supplementary cementitious materials on reduction of CO_2 emissions from concrete. Journal of Cleaner Production. 2015;**103**:774-783

[88] Tushar Q, Bhuiyan MA, Zhang G, Maqsood T, Tasmin T. Application of a harmonized life cycle assessment method for supplementary cementitious materials in structural concrete. Construction and Building Materials. 2022;**316**:125850

[89] Hossain MU, Poon CS, Dong YH, Xuan D. Evaluation of environmental impact distribution methods for supplementary cementitious materials. Renewable and Sustainable Energy Reviews. 2018;**82**:597-608

[90] Zhang Y, Zhang J, Luo W, Wang J, Shi J, Zhuang H, et al. Effect of compressive strength and chloride diffusion on life cycle CO_2 assessment of concrete containing supplementary cementitious materials. Journal of Cleaner Production. 2019;**218**:450-458

[91] Arrigoni A, Panesar DK, Duhamel M, Opher T, Saxe S, Posen ID, et al. Life cycle greenhouse gas emissions of concrete containing supplementary cementitious materials: Cut-off vs. substitution. Journal of Cleaner Production. 2020;**263**:121465

[92] Luukkonen T, Abdollahnejad Z, Yliniemi J, Kinnunen P, Illikainen M.

One-part alkali-activated materials: A review. Cement and Concrete Research. 2018;**103**:21-34

[93] Lu C, Zhang Z, Shi C, Li N, Jiao D, Yuan Q. Rheology of alkali-activated materials: A review. Cement Concrete Composition. 2021;**121**:104061

[94] Provis JL, Palomo A, Shi C. Advances in understanding alkali-activated materials. Cement and Concrete Research. 2015;**78**:110-125

[95] Oguntola O, Simske S. Continuous assessment of the environmental impact and economic viability of decarbonization improvements in cement production. Resources. 2023;**12**:95. DOI: 10.3390/resources12080095

[96] Robayo-Salazar R, Mejía-Arcila J, de Gutiérrez RM, Martínez E. Life cycle assessment (LCA) of an alkali-activated binary concrete based on natural volcanic pozzolan: A comparative analysis to OPC concrete. Construction and Building Materials. 2018;**176**:103-111

[97] Fernando S, Gunasekara C, Law DW, Nasvi M, Setunge S, Dissanayake R. Life cycle assessment and cost analysis of fly ash–rice husk ash blended alkali-activated concrete. Journal of Environmental Management. 2021;**295**:113140

[98] Adesanya E, Ohenoja K, Di Maria A, Kinnunen P, Illikainen M. Alternative alkali-activator from steel-making waste for one-part alkali-activated slag. Journal of Cleaner Production. 2020;**274**:123020

[99] Samarakoon M, Ranjith P, Duan WH, Haque A, Chen BK. Extensive use of waste glass in one-part alkali-activated materials: Towards sustainable construction practices. Waste Management. 2021;**130**:1-11

[100] Guo H, Shi C, Guan X, Zhu J, Ding Y, Ling T-C, et al. Durability of recycled aggregate concrete—A review. Cement Concrete Composition. 2018;**89**:251-259

[101] Thomas C, Setién J, Polanco J, Alaejos P, De Juan MS. Durability of recycled aggregate concrete. Construction and Building Materials. 2013;**40**:1054-1065

[102] Xing W, Tam VW, Le KN, Hao JL, Wang J. Life cycle assessment of recycled aggregate concrete on its environmental impacts: A critical review. Construction and Building Materials. 2022;**317**:125950

[103] Mah CM, Fujiwara T, Ho CS. Life cycle assessment and life cycle costing toward eco-efficiency concrete waste management in Malaysia. Journal of Cleaner Production. 2018;**172**:3415-3427

[104] Tan K-H, Wang T-Y, Zhou Z-H, Qin Y-H. Biochar as a partial cement replacement material for developing sustainable concrete: An overview. Journal of Materials in Civil Engineering. 2021;**33**:03121001

[105] Xiao J, Wang C, Ding T, Akbarnezhad A. A recycled aggregate concrete high-rise building: Structural performance and embodied carbon footprint. Journal of Cleaner Production. 2018;**199**:868-881

[106] Nanayakkara O, Gunasekara C, Sandanayake M, Law DW, Nguyen K, Xia J, et al. Alkali activated slag concrete incorporating recycled aggregate concrete: Long term performance and sustainability aspect. Construction and Building Materials. 2021;**271**:121512

[107] Alnahhal MF, Alengaram UJ, Jumaat MZ, Abutaha F, Alqedra MA, Nayaka RR. Assessment on engineering properties and CO_2 emissions of recycled

aggregate concrete incorporating waste products as supplements to Portland cement. Journal of Cleaner Production. 2018;**203**:822-835

[108] Shao Y, Mirza MS, Wu X. CO_2 sequestration using calcium-silicate concrete. Canadian Journal of Civil Engineering. 2006;**33**:776-784

[109] Xuan D, Zhan B, Poon CS. Assessment of mechanical properties of concrete incorporating carbonated recycled concrete aggregates. Cement Concrete Composition. 2016;**65**:67-74

[110] Dixit A, Du H, Dai PS. Carbon capture in ultra-high-performance concrete using pressurized CO_2 curing. Construction and Building Materials. 2021;**288**:123076

[111] Ravikumar D, Zhang D, Keoleian G, Miller S, Sick V, Li V. Carbon dioxide utilization in concrete curing or mixing might not produce a net climate benefit. Nature Communications. 2021;**12**:1-13

[112] Azzi ES, Karltun E, Sundberg C. Prospective life cycle assessment of large-scale biochar production and use for negative emissions in Stockholm. Environmental Science & Technology. 2019;**53**:8466-8476

[113] Aman AMN, Selvarajoo A, Lau TL, Chen W-H. Biochar as cement replacement to enhance concrete composite properties: A review. Energies. 2022;**15**:7662

[114] Gupta S, Kua HW, Low CY. Use of biochar as carbon sequestering additive in cement mortar. Cement Concrete Composition. 2018;**87**:110-129

[115] Gupta S, Kua HW. Factors determining the potential of biochar as a carbon capturing and sequestering construction material: Critical review. Journal of Materials in Civil Engineering. 2017;**29**:04017086

[116] Winters D, Boakye K, Simske S. Toward carbon-neutral concrete through biochar–cement–calcium carbonate composites: A critical review. Sustainability. 2022;**14**:4633

[117] Beatty DN, Williams SL, Srubar WV III. Biomineralized materials for sustainable and durable construction. Annual Review of Materials Research. 2022;**52**:411-439

[118] Rahbar N. Extending the life of self-healing structural materials. Matter. 2020;**2**:289-291

[119] Heveran CM, Williams SL, Qiu J, Artier J, Hubler MH, Cook SM, et al. Biomineralization and successive regeneration of engineered living building materials. Matter. 2020;**2**:481-494

[120] Srubar WV. Engineered living materials: Taxonomies and emerging trends. Trends in Biotechnology. 2021;**39**:574-583

[121] Stefanova A et al. Photosynthetic textile biocomposites: Using laboratory testing and digital fabrication to develop flexible living building materials. Science and Engineering of Composite Materials. 2021;**28**:223-236

[122] Sharma B, Singh A, Joshi S, Reddy MS. Utilization of sandstone waste in cement mortar for sustainable production of building materials through biomineralization. Journal of Sustainable Cement-Based Materials. 2023;**12**:712-720

[123] Delesky E, Jones R, Cook S, Cameron J, Hubler M, Srubar W III. Hydrogel-assisted self-healing of biomineralized living building

materials. Journal of Cleaner Production. 2023;**418**:138178

[124] Concrete Sustainable Council. The CO2-Module Technical Manual - What is the CO2-Module? 2024. Available from: https://csc.eco/2023/10/17/updated-fee-structure-for-2024/ [Accessed February 23, 2024]

Section 2

Research and Education Efforts for Reducing Carboon Footprint

Chapter 6

Perspective Chapter: Educational Technology under Scrutiny in Higher Education – A Framework for Balancing Environmental, Economic and Social Aspects in a Blended Design

Marieke Versteijlen and Marleen Janssen Groesbeek

Abstract

Following COVID-19, higher education (HE) has recognised the importance of educational technology (EdTech). With its growing influence on educational design, awareness of its role and impact on the sustainability transition in HE from an economic, environmental and social perspective is needed. Taking a holistic view, this chapter shows the opportunities and drawbacks of using EdTech in HE aiming at developing suggestions for responsible application. Economically, there are concerns because the earnings models of for-profit EdTech companies are focused on obtaining user data and benefit from a lack of transparency in data use, privacy and intellectual property. Environmentally, the use of EdTech can reduce the carbon footprint of HE institutions by opening a virtual space where students can learn and faculty can attend international conferences without commuting or travelling (by plane). However, device disposal and the energy consumption of hardware and data storage must be considered. Socially, using EdTech can foster the development of sustainability competencies if thoughtfully designed by applying pedagogical design principles for sustainability-oriented blended learning. Higher education can take control of a balanced use of EdTech in educational practice by focusing on ethical and human values and adopting a whole-institution approach to sustainability as included in the proposed framework.

Keywords: EdTech, sustainability, blended learning, student and staff travel, carbon footprint of higher education, EdTech earnings models

1. Introduction

Climate urgency requires everyone to reduce their emissions of greenhouse gas (GHG). Human activities have already evoked global warming with global surface temperature reaching 1.1°C above the temperature measured in 1850–1900 compared to

2011–2020 due to GHG emissions [1]. To mitigate climate change, the world needs a sustainable development approach aimed "to take the bold and transformative steps which are urgently needed to shift the world onto a sustainable and resilient path" ([2], p. 3).

Sustainable Development entails more than taking environmental measures. Already in 1987, the World Commission on Environment and Development (WCED) stated that "economics and ecology must be completely integrated into decision-making and lawmaking processes not just to protect the environment, but also to protect and promote development" [3] (Chapter II, 42) and also that environmental and economic problems are linked to social and political factors. Balancing economic, social and environmental factors served as a premise for the subsequent United Nations (UN) processes culminating in the Sustainable Development Goals (SDGs) [2].

In this chapter, we adopt the same premise for sustainability to study the affordances and drawbacks of educational technology (EdTech) in higher education (HE). EdTech companies are defined "as all the privately owned companies currently involved in the financing, production and distribution of commercial hardware, software, cultural goods, services and platforms for the educational market with the goal of turning a profit" ([4], p. 113). An important part of this educational market are the HE institutions.

1.1 Higher education

Higher education institutions are embedded in local communities linked with industry and business and are globally interconnected with other HE institutions so they have a wide stakeholder community. In many countries since the 1980s and 1990s, HE has been a major part of national strategies to achieve international competitive advantage in 'knowledge-based economies' [5]. As such, HE can play a major part in the sustainability challenges of this climate-changed world [6]. But HE's role of creating value for the knowledge economy has "gradually shifted to an ideal of the 'data-intensive' and 'digital-first university' of the twenty-first century that creates valuable new digital knowledge and develops digital data skills to support emerging capitalist data economies" ([5], p. 13). One can imagine this development is beneficial to EdTech companies or even encouraged by these (for-profit) companies. Still, the implications for a sustainable society of an academic environment infused with digital technology, data analytics and artificial intelligence are not clear.

Higher education is more than an institution creating and disseminating knowledge. Its purpose is to provide for 'good education'. In the context of sustainability, good education means acquiring the sustainability competencies needed to address the complex challenges this world faces creating a sustainable future. Competencies are a combination of knowledge, skills and attitudes, needed to accomplish the desired educational outcome [7]. HE should "bring students into a transformational relationship to professional and/or disciplinary knowledge that changes their sense of who they are, how they understand the world, and what they can do to change it" ([5], p. 72). So, this is more than learning digital competency (i.e. using information and communication technology (ICT) and digital devices to accomplish the desired outcome) to support the economy. Students should be well equipped to contribute to an economy that incorporates financial and non-financial values and strives for broad prosperity, i.e. acknowledging the importance of nature, environment, social cohesion, leisure time and happiness [8].

In the coming years, HE must find a balance in how to provide good education using digital technology to its own benefit and that of its students. There are promising opportunities.

1.2 Opportunities of digital technology

Using digital technology, students get a virtual space for their learning next to the physical space they already have when learning at their institution. In this virtual space, students have an abundance of knowledge at their disposal through the internet and they can interact or collaborate with their fellow students or anyone all over the world without the boundaries of time and place. These digital communication affordances may widen the horizon for students because perspectives of students from different disciplinary, national and cultural backgrounds can be taken into account when having group discussions and developing interpersonal or transboundary competencies [9, 10].

Place independence of learning may also provide environmental advantages. When organising on-campus education on fewer days per week, it may decrease the commute of students and campus facilities use [11, 12]. Considering the carbon footprint of a HE institution, the emissions due to student/staff travel and electricity purchased for campus facilities are the largest contributors [11, 13, 14]. On the other side, electronic equipment and infrastructure needed for educational technology also have an impact on the environment through the electricity consumed and the electronic waste (e-waste) generated [13].

In this chapter, we elaborate further on these issues. We balance the economic interests, the environmental dilemmas and the educational quality of EdTech, to end with a Framework for Sustainable Application of EdTech in HE and conclusions.

2. Balancing environmental, economic and educational factors of EdTech

2.1 EdTech and economic interests

The early 1990s are considered to be the beginning of EdTech with the involvement of large tech companies and influential private equity investors. Picciano and Spring [15] define it as the start of "The Great American Education-Industrial Complex". It is an ominous characterisation as an industrial complex is a socioeconomic concept wherein businesses become entwined in societal or political systems and institutions, creating or bolstering a profit economy from these systems. Such a complex is said to pursue its own interests regardless of, and often at the expense of, the best interests of society and individuals. This development intensified during and after the COVID-19 pandemic in which new networks, coalitions and alliances were established and the commercialisation and privatisation of education increased [16].

2.1.1 Influence of EdTech companies

The networks, coalitions and alliances of EdTech companies increase their influence on government and HE. An example of such an alliance is the collaboration in several projects between the Norwegian Centre for ICT in Education, an executive agency of the Norwegian Ministry of Education and Research, and the New Media

Consortium (NMC) [16]. This consortium was founded in 1993 by a group of hardware manufacturers, software developers and publishers. Although NMC has strong alliances with digital media and technology partners, such as Adobe, Apple, Intel and Pearson, their partnership with a governmental organisation was never under discussion. NMC's influence extends even further with the publication of the so-called Horizon Reports. These yearly reports give NMC's view on future technology trends in educational environments. For many HE institutions, this is a valuable resource for their educational technology procurement decisions [16]. In addition, academic articles often use publications from the Horizon Reports as background information. The 2020 EDUCAUSE Horizon Report mentions four future scenarios looking forward to the year 2030. In all four scenarios, digital technology plays a major role, transforming higher education into a more flexible, inclusive and personalised organisation, focused on teaching the skills the industry needs. In three of the four scenarios, this is at the expense of Humanities studies. In the most positive scenario Transformation, the student gets an artificial intelligence (AI) companion to "provide oversight, nudging, adaptive mentoring, research assistance, feedback on assignments, and friendly encouragement" (p. 36).

The EdTech marketing strategy is based on the certainty that technology will be needed in future for environmentally sustainable forms of teaching and learning. To make this proposition more attractive, EdTech entrepreneurs promise a 'technological fix' to societal problems facing higher education, such as the accessibility of education or having fewer resources for educational purposes due to budget cuts [4]. Particularly during the COVID-19 pandemic, the strategy of making alliances, problem-solving promises, predictions of needing educational technology and the 'emergency remote teaching' experiences [17] have not hurt EdTech companies. Because of the pandemic, the investments have grown exponentially. HolonIQ, an education market intelligence company, has calculated that the global venture capital investment in EdTech grew from US$ 500 million in 2010 to US$ 20 billion in 2021 [18]. Around 2020, research identified 20 EdTech unicorns in the world, of which 17 were spread between China and the United States (US) [19] and this number grew to 30 companies in 2023 with a combined value of over US$ 89 billion [20]. An unicorn is a private startup company valued at over US$ 1 billion. It is privately owned and not listed on a share market. That means that the transparency of its strategies and activities is limited. Because of the startup characteristics, the main goal of the private investors is to create the highest return on investment (ROI) in the shortest time possible.

2.1.2 Earnings models of EdTech companies

The platform model seems to have the greatest potential to increase profits for EdTech companies [18]. A digital educational platform creates a virtual space for teachers and students where information can be shared, resources for teaching and learning can be found and assignments can be uploaded. These platforms, also called learning management systems (LMSs), need regular upgrading with new features, can be integrated with other platform services and continuously extract digital data, creating a reliable revenue stream for investors [18]. Digital data gain value through processes like extracting meaningful information, enclosing it, storing, aggregating, analysing and transforming it into intelligence. This encompasses various sources, ranging from scholarly discussions in virtual learning environments to user behaviour data, including click-throughs on platforms, as well as metadata detailing users' devices, location and internet protocol addresses [19, 21].

The earnings model for for-profit companies is increasingly based on assetisation: creating assets that can be an endless means for returns. In practice, this results in "[charging] subscription fees for access to educational services and resources (monetary rent), integrating together in cross-platform ecosystems, and deriving value from user data through the creation of derivative products (data rent)" ([22], p. 6). Assetisation is not risk-free as privacy issues are just around the corner. The "use of deep neural networks within assetisation can by its very nature be not inspectable and thus inherently opaque. It involves complex, multi-stage decisions hard to scrutinise with human oversight" ([21], p. 21). There are multiple options for creating a return on student platform data. Student platform data might be sold to other companies for advertising purposes or, potentially, for enabling employers to identify and target students with promising competencies [22]. In Europe, the Digital Services Act is meant to protect online users' rights, filter illegal and harmful content, create new transparency for platform practices and ban targeted advertising to children [23]. Although the new regulations apply to countries that belong to the European Union (EU), it affects EdTech companies globally.

In addition to selling to external partners, data may also be used to improve educational services. Student platform data are the building blocks for Artificial Intelligence for EDucation (AIED) and learning analytics (LA). LA is "the measurement, collection, analysis and reporting of data about learners and their contexts, for purposes of understanding and optimising learning and the environments in which it occurs" ([24], p. 34). AI-based methods can turn student data into digital information to establish an individual and adaptive learning system based on the teaching and learning behaviour of the user [25]. As LA is used to improve the learning process, the purpose of AIED is to simulate and predict learning processes, think of making admission decisions, assessment and feedback, intelligent tutoring, profile and prediction of students dropping out [26].

Despite these promising opportunities, HE seems reluctant to implement LA or AIED solutions [25]. A constraining factor for the adoption of LA or AIED seems to be the advanced digital competencies a teacher needs to improve their teaching methods with AI-based functionalities [21, 25]. AI-based methods analyse the pattern of collected student platform data for developing the algorithms that can support the individual learner. These patterns may contain race, gender or other biases. In addition, the constant monitoring of students' digital behaviour on the platform increases surveillance and can amplify feelings of stress and anxiety around their loss of autonomy [5]. Williamson and Hogan [5] express their concern about this feeling of autonomy loss: "They are building robotised pedagogic environments in which key functions of teaching, such as observing student progress, providing feedback, scaffolding intellectual development, and assessment, are increasingly delegated to or augmented by automated AI technologies". These and other ethical concerns are acknowledged by international organisations, such as the United Nations Educational, Scientific and Cultural Organization (UNESCO). Although on 25 November 2021, 169 countries agreed on global standards for AI ethics, these are still in the recommendations stage [27, 28] and some EdTech companies complain about the vagueness of the proposed ethical principles [29]. Nguyen et al. [27] have gathered the ethical guidelines for AIED from the different reports and proposed several key ethical principles. One of these principles handles the need for transparency in data usage and accountability in AIED and LA:

> *"Principle of transparency in data and algorithms: The process of collecting, analysing, and reporting data should be transparent with informed consent and clarity of*

> *data ownership, accessibility, and the purposes for how data will be used. The AI algorithms should be explainable and justifiable for specific educational purposes."* *([27], p. 4229).*

The development of AI-based solutions becomes even more complex as there is not much generalisable, valid and reliable evidence gathered in research [25]. The European Union is not awaiting this research. In December 2023, the European Parliament reached a political agreement on the Artificial Intelligence Act. With these regulations, they aim to ensure that AI systems respect fundamental rights, safety and ethical principles [30].

Outsourcing course delivery to an online programme management (OPM) company is another way of making a financial return with HE EdTech products. These EdTech companies, for example, Coursera, share the revenue with university partners for massive open online course (MOOC)-related activities, advertising themselves "as having social missions to mobilise new technology to transform and improve not only the reach and quality of higher education but also the life and career prospects of EdTech-facilitated learners" ([31], p. 6). Mirrlees and Alvi [4] refer to MOOC-related activities as a way of making course delivery more efficient by "automating instruction". The learning content needs updating from time to time but online lectures can be watched multiple times. The reason this alliance is of interest to universities is that in countries worldwide the public sector, which includes education, is focused on efficiency, performance measurement, quality management, marketisation and accountability [5, 21].

Another aspect of the EdTech industry and its earnings model is the issue of intellectual property rights, particularly those of teachers working in a digital learning environment. LMS platforms contain the learning content (video lectures and teaching materials). It is unclear who has control and ownership over this content. If not regulated or contracted properly, what does that mean for academic freedom and content ownership, especially in countries with severe forms of internet censorship and surveillance? [5].

We have raised serious concerns about the involvement of EdTech companies in HE's online learning ambition. Technological innovations seem to precede the regulations needed to guarantee privacy, autonomy and transparency issues. The European Union takes the lead in filling this gap.

Higher Education and governments should adopt a data governance policy and "have a far stronger role in setting both the educational and regulatory agenda so that education serves the interests of a whole society: children, young people and life-learners, not just private capital investors" ([32], p. 91).

It is in the interest of society that students can develop digital competency [33]. They are the future human capital of a nation. Human capital refers to the economic value of an employee's professional competencies [33]. The use of EdTech is crucial in this development process, but HE's policy should be aimed at a human-centred approach.

2.2 EdTech and environmental dilemmas

Educational technology might play a role in decreasing the environmental impact of higher education activities by offering online communication possibilities. It creates opportunities for place-independent delivery of education or conference presentations. This may decrease commute-related travel and international business travel [14], and electricity purchased for campus facilities [11]. In this section, we discuss its impact on HEI's carbon footprint and the role of EdTech in adding to and reducing this carbon footprint.

2.2.1 Carbon emissions due to commute-related travel of students and staff

The environmental impact of a higher education institution (HEI) can be made transparent when measuring its carbon footprint. An internationally accepted accounting and reporting standard for companies and organisations is the Greenhouse Gas Protocol Initiative [34]. The GHG protocol classifies the emission sources into three scopes: scope 1: the direct emissions, scope 2: indirect emissions due to electricity purchased and scope 3: the indirect emissions as a consequence of HEI's activities. So, the electricity purchased for campus facilities belongs to scope 2 emissions, whereas the GHG emissions due to student and staff commute, and business travel belong to scope 3 emissions. Various studies acknowledge the significant proportion of student and staff travel, and the electricity purchased for campus facilities on HEI's carbon footprint [13].

Higher education students worldwide need to commute to their institutions 4 or 5 days per week. The amount of student travel emissions is dependent on how often students commute to attend classes on campus, the campus location and which travel mode they choose for travelling. The impact ranges between approximately 20 and 90 per cent of the overall carbon footprint [14, 35]. This broad range is due to dependencies on the accessibility of the campus, and the reliability and the very existence of the commute data [35].

How and which scope 3 emissions are measured is optional in the GHG protocol and in some studies, commute emissions are allocated to the individual carbon footprint of a student or staff member [36]. The reason why there might be the difficulty of obtaining reliable commute data. Data on commute behaviour can only be obtained by surveying students and staff and especially students are a difficult group to survey. Their limited response is often extrapolated to the whole university [37] or supplemented with parking data or registered student data where privacy issues come into play. Nevertheless, how difficult it may be, the commuting emissions are a direct consequence of HEI's activities, as students and staff often need to attend obligatory classes or are required to work from campus. Moreover, HEI can influence their commuting emissions by acting appropriately [38].

Most of these measures are directed at trying to change the travel mode of students and staff from carbon-intensive (i.e. single-occupancy petrol cars) to low-carbon modes (i.e. a bicycles or e-bikes) or promoting carpooling [39] and certainly, in car-dependent countries, such as Australia or the United States, this is a well-chosen strategy [39–41], but there are other options. The Intergovernmental Panel on Climate Change (IPCC) recommends, besides a modal shift, "avoiding journeys where possible—by, for example, (...) utilizing advanced information and communication technologies (ICT)" ([42], p. 603). In HE, EdTech can offer possibilities to decrease student and staff commute [12]. These possibilities relate to the use of digital technologies to enable location-independent learning.

2.2.2 Carbon emissions due to business travel

Academic staff also need to commute but their main contribution to the carbon footprint are their visits to international conferences and seminars, often by plane [43]. These business travels can account for about 3–55 per cent of a HEI's footprint.

Digital technologies can play a role in mitigating the large amount of carbon emissions due to these meetings [44]. While one can dispute whether the digital technology required should be subsumed under the banner of EdTech, it is strongly interrelated and contributing to HE's carbon footprint. It seems crucial for academics to keep abreast of the latest developments in their field and network in person at these events [45]. The impact and quality of research are dependent on international collaboration and these collaborations are usually initiated and sustained through physical meetings [46].

Although technically possible, as proven during the COVID-19 pandemic, a transition to fully online conferences does not seem to be a desirable solution to the problem of high-carbon transportation for attending. Nevertheless, organising these international meetings should focus on reducing carbon emissions due to travel and there are opportunities enhanced by digital technology, such as hybrid conferences [43, 47]. A large contribution to the total emissions of an international conference is due to a relatively small group of participants [47]. They could present their research and take part in discussions using remote conferencing services. Another possibility is to organise the hybrid conference at two different venues, although this may have the rebound effect that more academics will attend the conference [48]. Both solutions roughly halve travel-related emissions [47, 49].

2.2.3 Energy consumption of EdTech use

Although contributing to decreasing education-related travel emissions, EdTech services are not carbon-neutral. They generate GHG emissions due to the electricity needed to power devices. They provide the necessary video- and audio transfer, and storage data. The volume of these emissions is nation-dependent because electricity can be generated in various ways. Data from the European Environmental Agency (EEA) show that in 2020 producing 1 kWh of electricity in France emitted ca. 60 g of carbon dioxide (CO_2), whereas producing the same amount in Estonia led to 621 g of CO_2 emissions [50]. To evaluate the use of EdTech to decrease the HEI's carbon footprint, we should compare the carbon impact of student and staff travel with the impact of EdTech use.

Regarding location-independent learning, Caird et al. [11] compared energy consumption with associated carbon emissions generated by distance-based HE teaching models with HE on-campus models. Their findings indicate that distance-based models achieve a carbon reduction of 83 per cent in comparison with on-campus models, mainly due to commute-related student travel and campus site operations [11]. The COVID-19 pandemic forced HE in countries categorised as developed economies to an immediate transition from on-campus to an online environment from home [51]. Some researchers took this opportunity to study the carbon impact of this online learning environment. There are two contradictory studies regarding the carbon impact of the increased use of energy devices needed to study or work from home. Filimonau et al. [52] stated that the carbon impact of devices used for online education was almost equal to that of staff and student commute at Bournemouth University in the United Kingdom (UK), and in China at Wuhan University, Yin et al. [53] concluded that online education can significantly reduce energy consumption and lower carbon emissions when considering transportation and electricity consumption [52, 53]. It should be noted that Filimonau et al. [52] assumed that staff and students used their devices for study or work at home 8 hours a day and 5 days per week which is, in our opinion, highly unlikely.

Clearly, more research is needed with reliable data on student travel and the additional energy required during home education. Still, the study of Yin et al. [53] confirms the results of Caird et al. [11] that online education can significantly decrease energy consumption and associated carbon emissions. HE, in their quest for carbon neutrality, could take advantage of these findings by organising education in a hybrid form (the so-called blended learning), substituting on-campus learning with online learning a few days per week. This would reduce the days students have to commute to their institution to attend classes.

2.2.4 Energy consumption and disposal issues of EdTech

In Section 2, we saw that in future, HE will probably increase the application of AIED and LA. Developing an AI model for AIED involves training deep learning models on vast amounts of data. This consumes a significant amount of energy during both training and validation and these computations have a large carbon footprint [54]. Another issue is that AIED and LA require extensive datasets, usually stored on servers in data centres. These servers not only consume a vast amount of energy to store data but also respond to data requests. Besides energy, water is extensively used for liquid cooling in these data centres [55].

The last issue we need to address in relation to EdTech is the disposal of electronic devices, adding to the generation of electronic waste (e-waste). In 2019, 53.6 million metric tonnes of e-waste was generated and only 17.4 per cent was collected and recycled [56]. E-waste is a health and environmental risk and contains toxic additives or harmful substances such as mercury, which damages the human brain and/or coordination system [56].

The environmental gains of using EdTech to decrease education-related travel (student and staff commute and business travel) are accompanied by concerns about energy consumption and e-waste disposal. Considering HE's carbon footprint, it seems that decreasing education-related travel outweighs the energy consumption of the devices [11, 14, 53]. Nevertheless, the environmental costs of the development and use of AIED and LA should be part of an environmentally aware approach to EdTech [57].

2.3 EdTech and educational quality

Educational technology will likely play a more significant role in HE shortly, given the impact of the COVID-19 pandemic on HE and the increasing influence of EdTech on government and educational policy, The question remains how will EdTech fulfil this highly influential role? How will they pay attention to all their responsibilities: ethically and sustainably?

As we have seen in the former section, EdTech may be used to lower HE's carbon footprint if learning activities are scheduled with a clear division between on-campus and online learning days. Thus, organising education can reduce students' commute to attend on-campus classes. This educational model is called blended learning [58]. As the mission of HE is to provide 'good education', blended learning should not compromise or better still, enhance educational quality.

2.3.1 Quality assurance of blended learning

Quality assurance is a complex issue in HE. Good education is difficult to quantify, for instance, measuring the quality of a graduate student might be done by looking at their employability. But how does one measure and assess if this is due to the education this student has received [59]? Focusing on educational outcomes and performance seems hazardous for quality assurance but focusing on the learning and teaching process of how these outcomes are obtained seems to be a better approach. The E-xcellence framework describes benchmarks for quality assurance of e-learning, which are grouped into six key areas: strategic management, curriculum design, course design, course delivery, staff and student support [60].

During the COVID-19 pandemic, the rapid transition to online education in HE revealed the deficiencies in these key areas. During online course delivery, problems arose with infrastructure and availability of devices just as with lecturers' pedagogic knowledge and experience [61]. This resulted in so-called "emergency remote teaching" [62]. Several problems are caused by the sudden transition and can be solved in time, but some are more difficult to tackle—for example, accessibility to online learning services, a quiet study environment or the well-being of students. Accessibility and study environment are dependent on students' social-economic status. Students with a low socioeconomic background usually need HEI's environment for the use of computers and free and fast internet to study online, undisturbed by human and pet intrusions [63, 64]. Not having access to these facilities and experiencing technological problems at home during COVID-19, aroused feelings of anxiety, anger and boredom due to not being able to hear and take the entire online class [65]. These negative emotions were also triggered by increased workload [66] and insufficient digital competency [65]. To a large extent, the severity of these problems was caused by the sudden transition to online learning but also in a carefully planned and designed blended learning unit these accessibility issues and failing technological abilities of faculty and students should be addressed.

2.3.2 Pedagogical design principles for sustainability-oriented blended learning

Zooming in on the design of blended learning, the pedagogical quality of education should not be compromised. As we live in a climate-changed world with complex sustainability challenges, students must be "capable of analysing, evaluating, and synthesizing complex issues and of applying learning in new contexts. At the same time, they appreciate the advantages of cooperating to pursue a common goal" [67]. Transformation to a sustainable world needs 'change agents', who are aware of what they can or want to change and know how to take action for implementation [68]. This requires a pedagogical approach that supports students in developing sustainability competencies. These competencies provide the students with the necessary knowledge and skills to analyse systems across different domains, anticipate future challenges, apply ethical values, design and implement transformative interventions and engage stakeholders in the process [68, 69]. The pedagogical design principles of Versteijlen and Wals [10] can be used to guide the design and assess the pedagogical quality of a sustainability-oriented blended learning unit. This is a learning unit organised with a clear division between on-campus and online learning days to decrease student commute and a sustainability-oriented pedagogical approach to learning. This requires a thoughtful design because it is a challenge to combine virtual and physical space in such a way that the strengths of both are exploited and complement each other.

Table 1 shows the usability of EdTech for designing sustainability-oriented blended learning by connecting the design principles of Versteijlen and Wals [10] with their relevance for learning and how EdTech can be supportive. The first two

Design principle	Relevance to learning	EdTech support
Aiming at self-regulation and self-awareness of learning and practice in the student's learning process	In blended learning, students have more agency over where and when they study. Digital tools can stimulate the development of the necessary self-regulating skills. A reflective environment with digital and physical communication possibilities promotes creating self-awareness of one's own values	24/7 access to the digital environment, personalised learning plans, learning management tools, online tests and quizzes, online reflective journals, virtual and augmented reality
Fostering a safe and social learning environment	Pointing to a feeling of freedom to express oneself without being afraid of failure is a prerequisite of creating self-awareness. This is best encouraged by the lecturer in the physical space and reinforced in the virtual space.	Communication platforms, social networking, global community tools, video feedback and virtual coffee shop
Facilitating (a)synchronous interaction and discussion among fellow students and with the lecturer to stimulate reflection and critical thinking	Different approaches to learning can be taken into account by offering synchronous as well as asynchronous discussions of topics or feedback.	Online debate platform, web annotation tool, video discussion platform, online tutoring, brainstorming tool
Transforming learning through acquisition and inquiry into an active process based on existing knowledge in which new knowledge is constructed to contribute to sustainability	It motivates students to acquire knowledge through their preferred (online) content representation. Inquiry stimulates system thinking competency. Data are easily accessible using the internet.	Multimedia content, online lectures, digital (formative) tests, virtual classroom, data collection, search engines, interactive tutorials, virtual and augmented reality
Working on authentic and action-oriented tasks with scaffolded and theory-based practice meeting the learning preferences of students	Transfer of learning from the classroom to the professional setting by designing online learning tasks and tools that align with the workplace setting while critically examining sustainability opportunities.	Peer feedback tools, interactive learning, virtual classroom, serious games, lab simulations, personalised assignments
Inter/transdisciplinary collaboration for constructing a shared outcome through participation and negotiation with fellow students in a technologically enhanced learning environment	Developing interpersonal competency through incorporating different perspectives in group discussions for addressing complex social, ecological, technical and other problems to bring about transformative change	Integrated team collaboration, project organisation, digital boards for organising and sharing content, wiki platform.

Table 1.
Usability of EdTech for designing sustainability-oriented blended learning. Based on the pedagogical design principles for sustainability-oriented blended learning [10, 70].

design principles are more general in nature and supportive of the last four design principles, which are about learning activities in which students interact and discuss, acquire knowledge, bring theoretical knowledge into practice and collaborate incorporating different perspectives [10, 70]. In a blended design, these learning activities should be interwoven according to the Conversational Framework of Laurillard [71]. This framework specifies the iterative reflective interactions between student-lecturer and student-fellow students on two contrasting levels: 1. articulating and discussing theory and 2. experimenting and practising on goal-oriented tasks [71].

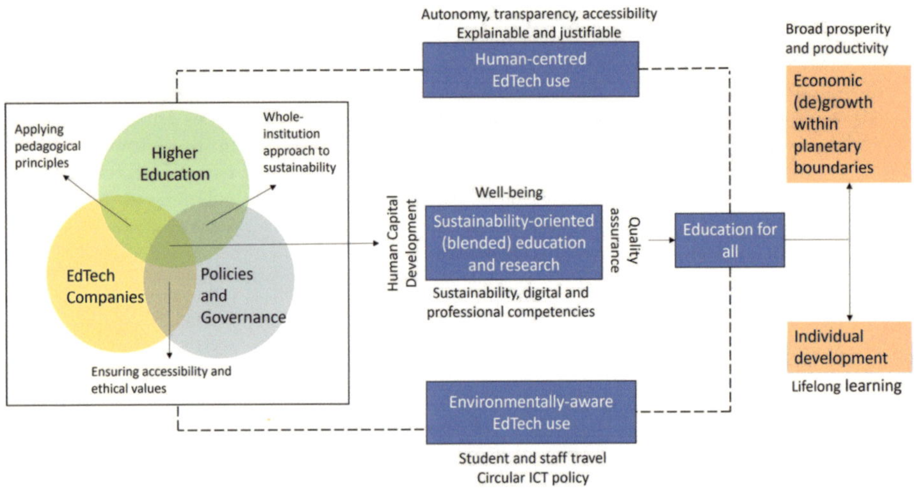

Figure 1.
Framework for sustainable EdTech application in higher education. Based on the framework for a humanist ecosystem of higher education [33].

In this section, we discussed whether EdTech in a blended learning design would compromise or enhance educational quality in HE. Provided certain conditions are met, we can say that EdTech improves educational quality and is indispensable in a design for blended learning. These conditions involve the benchmarks for quality assurance of e-learning and measures to ensure the accessibility to learning and well-being of students. The relationship between HE, EdTech companies and their policies and governance should support the development of human capital this climate-changed world needs.

2.4 A sustainability-oriented perspective on EdTech in higher education

With sustainability in mind, we have balanced the drawbacks and affordances of EdTech in HE. In **Figure 1**, the key concepts of EdTech under scrutiny are depicted in a Framework for Sustainable EdTech Application in higher education. This framework builds upon the framework for a humanist ecosystem of higher education given in Sharma [33], but we adopt a sustainability-oriented perspective on the role of EdTech in the education system. This system is represented by the relationship between EdTech companies, HE and their governance and policies. As stated before, HE plays a crucial role in the sustainability challenges facing the world. This responsibility requires higher education to embed sustainability in all aspects of its policies, a so-called "whole-institution approach" [72]. These aspects relate to education, research, own sustainability-related behavioural practices, community engagement, leadership and ethos, and professional development of staff [73]. We can add EdTech to these aspects. This approach ensures sustainability-oriented (blended) education and research, organised in such a way that the deployment of EdTech is human-centred and environmentally aware.

3. Conclusion

Educational Technology has become an integral part of the (higher) education system and improves education by offering students a virtual space next to the physical one where they can learn and interact with lecturers and fellow students. This transition has perhaps been accelerated by the COVID-19 pandemic and so many venture capital investors see the market potential of HE and its students who have grown used to being online continuously.

The promises and predictions of Edtech seem to be attractive, but HE institutions have to be vigilant. In this chapter, we have identified being in control and privacy issues, educational quality and sustainability focal points to which we have to pay attention when applying EdTech in HE. In this chapter, we approached EdTech from a sustainable development perspective, balancing economic, environmental and educational factors.

There is a tension between the economic interests of the for-profit EdTech companies, private owners and educational organisations. We have to be on the alert whether the EdTech companies will deliver on the promises they make or even whether those promises and predictions will get in the way of good education. The EdTech companies promise flexible and personalised education but fail to mention that EdTech is not accessible to everyone, can come at the expense of lecturer's and student's autonomy and raises serious privacy issues. HE's and EdTech companies' policies should ensure that everyone has access to their facilities and that ethical values are upheld. In addition, clarity in data usage and ownership is needed when scrutinising their earnings model of assetisation. This could lead to lower ROIs for the private investors in the EdTech companies, but it may remove some of the reservations HE has about introducing LA and AIED in their educational practices.

Considering the carbon footprint of the HEIs, one of the main sources of carbon emissions is the education-related travel of students and staff. The additional virtual space by using EdTech offers opportunities for organising education in such a way that there is a clear division between on-campus and online learning days during a week. This may reduce student commute. In addition, virtual space can be used during international conferences, decreasing business travel of staff. Still, a thoughtful mix of virtual and physical space is needed because physical encounters still add value to teaching and research. In addition, energy consumption and e-waste issues of EdTech should be considered.

A thoughtful mix can be realised in educational practice by applying the pedagogical design principles for sustainability-oriented blended learning. The experiences of online learning during COVID-19 showed that students' emotional well-being and self-regulation were an issue but they welcomed the additional time freed up by not having the daily commute [70]. A thoughtful blended design offers 'the best of both worlds' and probably will improve educational quality.

The developed Framework for Sustainable EdTech Application in HE contains the most significant points when applying EdTech in educational practice. Some of the ethical issues can be resolved by acknowledging and applying standards for AI ethics [27]. However, HEIs and governments must remain vigilant on privacy and other ethical issues. EdTech use can be supportive of educational quality and sustainable organisation, but HE and Edtech companies must ensure in cooperation that human values and well-being are guaranteed.

Conflict of interest

The authors declare no conflict of interest.

Author details

Marieke Versteijlen* and Marleen Janssen Groesbeek
Avans University of Applied Sciences, Breda, The Netherlands

*Address all correspondence to: mbc.versteijlen@avans.nl

References

[1] IPCC. Climate Change 2023 [Internet]. 2023. Available from: https://www.ipcc.ch/report/sixth-assessment-report-cycle/ [Accessed: February 27, 2024]

[2] United Nations. Transforming Our World: The 2030 Agenda for Sustainable Development [Internet]. 2015. Available from: https://sdgs.un.org/publications/transforming-our-world-2030-agenda-sustainable-development-17981 [Accessed: February 27, 2024]

[3] Brundtland G. Report of the World Commission on Environment and Development: Our Common Future. United Nations General Assembly Document A/42/427; 1987

[4] Mirrlees T, Alvi S. EdTech Inc.: Selling, Automating and Globalizing Higher Education in the Digital Age. New York: Routledge; 2019. DOI: 10.4324/9780429343940

[5] Williamson B, Hogan A. Pandemic Privatisation in Higher Education: Edtech and University Reform. Brussels, Belgium: Education International; 2021

[6] Purcell WM, Henriksen H, Spengler JD. Universities as the engine of transformational sustainability toward delivering the sustainable development goals: "Living labs" for sustainability. International Journal of Sustainability in Higher Education. 2019;**20**(8):1343-1357. DOI: 10.1108/IJSHE-02-2019-0103

[7] Lozano R, Barreiro-Gen M, Lozano FJ, Sammalisto K. Teaching sustainability in European higher education institutions: Assessing the connections between competences and pedagogical approaches. Sustainability. 2019;**11**:1602. DOI: 10.3390/su11061602

[8] Bijl R. Never waste a good crisis: Towards social sustainable development. Social Indicators Research. 2011;**102**:157-168. DOI: 10.1007/s11205-010-9736-y

[9] De Kraker J, Corvers R, Lansu A. E-learning for sustainable development: Linking virtual mobility and transboundary competence development. In: Azeiteiro U, Leal Filho W, Caeiro S, editors. E-learning and Education for Sustainability. New York: Peter Lang Publishing Group; 2014. pp. 29-46

[10] Versteijlen M, Wals AE. Developing design principles for sustainability-oriented blended learning in higher education. Sustainability. 2023;**15**:8150. DOI: 10.3390/su15108150

[11] Caird S, Lane A, Swithenby E, Roy R, Potter S. Design of higher education teaching models and carbon impacts. International Journal of Sustainability in Higher Education. 2015;**16**:96-111

[12] Versteijlen M, van Wee B, Wals A. Exploring sustainable student travel behaviour in The Netherlands: Balancing online and on-campus learning. International Journal of Sustainability in Higher Education. 2021;**22**(8):146-166. DOI: 10.1108/IJSHE-10-2020-0400

[13] Valls-Val K, Bovea MD. Carbon footprint in Higher Education Institutions: A literature review and prospects for future research. Clean Technologies and Environmental Policy. 2021;**23**:2523-2542

[14] Versteijlen M, Salgado FP, Groesbeek MJ, Counotte A. Pros and cons of online education as a measure to reduce carbon emissions in higher education in the Netherlands. Current

Opinion in Environmental Sustainability. 2017;**28**:80-89. DOI: 10.1016/j.cosust.2017.09.004

[15] Picciano AG, Spring J. The Great American Education-Industrial Complex: Ideology, Technology, and Profit. New York: Routledge; 2013. DOI: 10.4324/9780203120613

[16] Haugsbakk G. Technology giants, educational policy and a preliminary mapping of networks and channels of influence in a Norwegian context. Seminar Net. 2021;**17**(2). DOI: 10.7577/seminar.4303

[17] Marinoni G, Van't Land H, Jensen T. The Impact of Covid-19 on Higher Education around the World. Paris: International Association of Universities; 2020

[18] Davies H, Eynon R, Komljenovic J, Williamson B. Investigating the financial power brokers behind EdTech. In: Livingstone S, Pothong K, editors. Education Data Futures: Critical, Regulatory and Practical Reflections. Digital Futures Commission, 5Rights Foundation; 2022. Available from: https://educationdatafutures.digitalfuturescommission.org.uk/essays/competing-interests-in-education-data/investigation-financial-power-brokers-edtech [Accessed: February 27, 2024]

[19] Komljenovic J. The future of value in digitalised higher education: Why data privacy should not be our biggest concern. Higher Education. 2022;**83**:119-135. DOI: 10.1007/s10734-020-00639-7

[20] Shustova I, Polevaya E, Ivanov A. The impact of Edtech: Strengths and opportunities. E3S Web of Conferences. 2023;**458**:06012. DOI: 10.1051/e3sconf/202345806012

[21] de Andrade LHA, Thomas DA, Laterza V. The rise of EdTech platforms in higher education: Mapping themes from emerging critical literature. In: Pinheiro R, Edelhard Tømte C, Barman L, Degn L, Geschwind L, editors. Digital Transformations in Nordic Higher Education. London: Palgrave Macmillan; 2023. pp. 27-51. DOI: 10.1007/978-3-031-27758-0_2

[22] Williamson B, Komljenovic J. Investing in imagined digital futures: The techno-financial 'futuring' of edtech investors in higher education. Critical Studies in Education. 2023;**64**:234-249. DOI: 10.1080/17508487.2022.2081587

[23] European Commission. The Digital Services Act [Internet]. 2024. Available from: https://commission.europa.eu/strategy-and-policy/priorities-2019-2024/europe-fit-digital-age/digital-services-act_en [Accessed: February 27, 2024]

[24] Siemens G, Long P. Penetrating the fog: Analytics in learning and education. Educause Review. 2011;**46**:30

[25] Renz A, Hilbig R. Prerequisites for artificial intelligence in further education: Identification of drivers, barriers, and business models of educational technology companies. International Journal of Educational Technology in Higher Education. 2020;**17**:1-21. DOI: 10.1186/s41239-020-00193-3

[26] Rienties B, Køhler Simonsen H, Herodotou C. Defining the boundaries between artificial intelligence in education, computer-supported collaborative learning, educational data mining, and learning analytics: A need for coherence. Frontiers in Education. 2020;**5**:128. DOI: 10.3389/feduc.2020.00128

[27] Nguyen A, Ngo HN, Hong Y, Dang B, Nguyen B-PT. Ethical principles for artificial intelligence in education. Education and Information Technologies.

2023;**28**:4221-4241. DOI: 10.1007/s10639-022-11316-w

[28] UNESCO. Recommendation on the Ethics of Artificial Intelligence. Paris: UNESCO; 2022. Available from: https://unesdoc.unesco.org/ark:/48223/pf0000381137 [Accessed: February 27, 2024]

[29] Kousa P, Niemi H. AI ethics and learning: EdTech companies' challenges and solutions. Interactive Learning Environments. 2022;**31**(10):6735-6746. DOI: 10.1080/10494820.2022.2043908

[30] European Commission. Shaping Europe's Digital Future [Internet]. 2024. Available from: https://digital-strategy.ec.europa.eu/en/policies/regulatory-framework-ai [Accessed: February 27, 2024]

[31] Thomas DA, Nedeva M. Broad online learning EdTech and USA universities: Symbiotic relationships in a post-MOOC world. Studies in Higher Education. 2018;**43**:1730-1749. DOI: 10.1080/03075079.2018.1520415

[32] Davies J. An analysis of the sustainability of different methods of delivering higher education. In: Leal Filho W, Brandli L, Kuznetsova O, Paço A, editors. Integrative Approaches to Sustainable Development at University Level. World Sustainability Series. Berlin: Springer; 2015. pp. 67-79. DOI: 10.1007/978-3-319-10690-8_5

[33] Sharma H. Mapping the global EdTech revolution during the pandemic: From 'determinism' to 'solutionism'. In: Mogaji E, Jain V, Maringe F, Hinson RE, editors. Re-imagining Educational Futures in Developing Countries. London: Palgrave Macmillan; 2022. pp. 119-137. DOI: 10.1007/978-3-030-88234-1_7

[34] WBCSD. Setting Operational Boundaries. USA: World Resources Institute and World Business Council for Sustainable Development; 2014. pp. 25-33

[35] Helmers E, Chang CC, Dauwels J. Carbon footprinting of universities worldwide: Part I—Objective comparison by standardized metrics. Environmental Sciences Europe. 2021;**33**:1-25. DOI: 10.1186/s12302-021-00454-6

[36] Townsend J, Barrett J. Exploring the applications of carbon footprinting towards sustainability at a UK university: Reporting and decision making. Journal of Cleaner Production. 2015;**107**:164-176. Available from: http://www.sciencedirect.com/science/article/pii/S0959652613007634

[37] Ozawa-Meida L, Brockway P, Letten K, Davies J, Fleming P. Measuring carbon performance in a UK University through a consumption-based carbon footprint: De Montfort University case study. Journal of Cleaner Production. 2013;**56**:185-198. Available from: http://www.sciencedirect.com/science/article/pii/S0959652611003593

[38] Kiehle J, Kopsakangas-Savolainen M, Hilli M, Pongrácz E. Carbon footprint at institutions of higher education: The case of the University of Oulu. Journal of Environmental Management. 2023;**329**:117056. DOI: 10.1016/j.jenvman.2022.117056

[39] Sen G, Chau H-W, Tariq MAUR, Muttil N, Ng AW. Achieving sustainability and carbon neutrality in higher education institutions: A review. Sustainability. 2021;**14**:222. DOI: 10.3390/su14010222

[40] Hancock L, Nuttman S. Engaging higher education institutions in the

challenge of sustainability: Sustainable transport as a catalyst for action. Journal of Cleaner Production. 2014;**62**:62-71. DOI: 10.1016/j.jclepro.2013.07.062

[41] Zhou JP. Sustainable commute in a Car-dominant city: Factors affecting alternative mode choices among university students. Transportation Research Part A: Policy and Practice. 2012;**46**:1013-1029. DOI: 10.1016/j.tra.2012.04.001

[42] IPCC. Climate Change 2014 Mitigation of Climate Change. Cambridge: Cambridge University Press; 2014. pp. 599-650. Available from: https://www.ipcc.ch/report/ar5/wg3/ [Accessed: February 27, 2024]

[43] Leochico CFD, Di Giusto ML, Mitre R. Impact of scientific conferences on climate change and how to make them eco-friendly and inclusive: A scoping review. The Journal of Climate Change and Health. 2021;**4**:100042. DOI: 10.1016/j.joclim.2021.100042

[44] Jäckle S. Reducing the carbon footprint of academic conferences by online participation: The case of the 2020 virtual European consortium for political research general conference. PS: Political Science and Politics. 2021;**54**:456-461. DOI: 10.1017/S1049096521000020

[45] Eriksson E, Söderberg MW, Wormbs N. Exceptionalism and evasion: How scholars reason about air travel. In: Bjørkdahl K, Franco Duharte AS, editors. Academic Flying and the Means of Communication. Singapore: Palgrave Macmillan; 2022. pp. 159-183. DOI: 10.1007/978-981-16-4911-0_7

[46] Sugimoto CR, Robinson-García N, Murray DS, Yegros-Yegros A, Costas R, Larivière V. Scientists have most impact when they're free to move. Nature. 2017;**550**:29-31. DOI: 10.1038/550029a

[47] Jäckle S. The carbon footprint of travelling to international academic conferences and options to minimise it. In: Bjørkdahl K, Franco Duharte AS, editors. Academic Flying and the Means of Communication. Singapore: Palgrave Macmillan; 2022. pp. 19-52. DOI: 10.1007/978-981-16-4911-0_2

[48] Coroama VC, Hilty LM, Birtel M. Effects of Internet-based multiple-site conferences on greenhouse gas emissions. Telematics and Informatics. 2012;**29**:362-374. Available from: http://www.sciencedirect.com/science/article/pii/S0736585311000773

[49] Coroama VC, Moberg Å, Hilty LM. Dematerialization through electronic media? In: Hilty L, Aebischer B, editors. ICT Innovations for Sustainability. Advances in Intelligent Systems and Computing. Vol. 310. Berlin: Springer; 2015. pp. 405-421. DOI: 10.1007/978-3-319-09228-7_24

[50] European Environmental Agency. CO2-Emission Intensity From Electricity Generation [Internet]. 2022. Available from: Greenhouse Gas Emission Intensity of Electricity Generation—European Environment Agency (europa.eu) [Accessed: February 27, 2024]

[51] Crawford J, Butler-Henderson K, Rudolph J, Malkawi B, Glowatz M, Burton R, et al. COVID-19: 20 countries' higher education intra-period digital pedagogy responses. Journal of Applied Learning and Teaching. 2020;**3**:1-20. DOI: 10.37074/jalt.2020.3.1.7

[52] Filimonau V, Archer D, Bellamy L, Smith N, Wintrip R. The carbon footprint of a UK University during the COVID-19 lockdown. Science of the Total Environment. 2021;**756**:143964. DOI: 10.1016/j.scitotenv.2020.143964

[53] Yin Z, Jiang X, Lin S, Liu J. The impact of online education on carbon emissions in the context of the COVID-19 pandemic—Taking Chinese universities as examples. Applied Energy. 2022;**314**:118875. DOI: 10.1016/j.apenergy.2022.118875

[54] Salehi S, Schmeink A. Data-centric green artificial intelligence: A survey. IEEE Transactions on Artificial Intelligence. 2023;**1**(1):1-18. DOI: 10.1109/TAI.2023.3315272

[55] Al Kez D, Foley AM, Laverty D, Del Rio DF, Sovacool B. Exploring the sustainability challenges facing digitalization and internet data centers. Journal of Cleaner Production. 2022;**371**:133633. DOI: 10.1016/j.jclepro.2022.133633

[56] Forti G, Baldé CP, Kuehr R, Bel G. The Global E-Waste Monitor 2020: Quantities, Flows and the Circular Economy Potential. Bonn, Geneva, Rotterdam: United Nations University (UNU)/United Nations Institute for Training and Research (UNITAR); 2020

[57] Selwyn N. On the limits of artificial intelligence (AI) in education. Nordisk Tidsskrift for Pedagogikk og Kritikk. 2024;**10**:3-14. DOI: 10.23865/ntpk.v10.6062

[58] Allen IE, Seaman J. Sizing the Opportunity: The Quality and Extent of Online Education in the United States, 2002 and 2003. Newburyport: Sloan Consortium (NJ1); 2003

[59] Ellis RA, Calvo RA. Minimum indicators to assure quality of LMS-supported blended learning. Educational Technology & Society. 2007;**10**:60-70

[60] Kear K, Williams K, Rosewell J. Excellence in e-learning: A quality enhancement approach. In: EFQUEL Innovation Forum 2014/International LINQ Conference 2014: Changing the Trajectory - Quality for Opening Up Education. Crete; 7-9 May 2014

[61] Ali W. Online and remote learning in higher education institutes: A necessity in light of COVID-19 pandemic. Higher Education Studies. 2020;**10**:16

[62] Hodges CB, Moore S, Lockee BB, Trust T, Bond MA. The Difference Between Emergency Remote Teaching and Online Learning. Educause; 2020. Available from: https://er.educause.edu/articles/2020/3/the-difference-between-emergency-remote-teaching-and-online-learning [Accessed: April 9, 2024]

[63] Adedoyin OB, Soykan E. Covid-19 pandemic and online learning: The challenges and opportunities. Interactive Learning Environments. 2023;**31**:863-875. DOI: 10.1080/10494820.2020.1813180

[64] Cramarenco RE, Burcă-Voicu MI, Dabija D-C. Student perceptions of online education and digital technologies during the COVID-19 pandemic: A systematic review. Electronics. 2023;**12**:319. DOI: 10.3390/electronics12020319

[65] Acosta-Gonzaga E, Ruiz-Ledesma EF. Students' emotions and engagement in the emerging hybrid learning environment during the COVID-19 pandemic. Sustainability. 2022;**14**:10236. DOI: 10.3390/su141610236

[66] Raccanello D, Balbontín-Alvarado R, da Silva BD, Burro R, Cheraghi M, Dobrowolska B, et al. Higher education students' achievement emotions and their antecedents in e-learning amid COVID-19 pandemic: A multi-country survey. Learning and Instruction. 2022;**80**:101629. DOI: 10.1016/j.learninstruc.2022.101629

[67] UNESCO. SDG Resources for Educators—Quality Education [Internet]. 2021. Available from: https://en.unesco.org/themes/education/sdgs/material/04#secondary [Accessed: February 27, 2024]

[68] Brundiers K, Barth M, Cebrián G, Cohen M, Diaz L, Doucette-Remington S, et al. Key competencies in sustainability in higher education—Toward an agreed-upon reference framework. Sustainability Science. 2021;**16**:13-29. DOI: 10.1007/s11625-020-00838-2

[69] Wiek A, Withycombe L, Redman CL. Key competencies in sustainability: A reference framework for academic program development. Sustainability Science. 2011;**6**:203-218. DOI: 10.1007/s11625-011-0132-6

[70] Versteijlen M, Wals AE, van Wee B. Design and evaluation of a sustainable blended study programme in higher education. Frontiers in Education. 2023;**8**. DOI: 10.3389/feduc.2023.1222962

[71] Laurillard D. Teaching as a Design Science: Building Pedagogical Patterns for Learning and Technology. New York: Routledge; 2013

[72] Kohl K, Hopkins C, Barth M, Michelsen G, Dlouhá J, Razak DA, et al. A whole-institution approach towards sustainability: A crucial aspect of higher education's individual and collective engagement with the SDGs and beyond. International Journal of Sustainability in Higher Education. 2021;**23**(2):218-236. DOI: 10.1108/IJSHE-10-2020-0398

[73] Mathie RG, Wals A. Whole School Approaches to Sustainability: Exemplary Practices from around the World. Wageningen: Wageningen University, Education & Learning Sciences; 2022. DOI: 10.18174/566782

Chapter 7

Perspective Chapter: Unlocking Energy and Natural Resources for Sustainable Development in Nigeria – The Role of Energy Research Centers

Emmanuel Ogbomida, Abdullahi Mustapha, Chukwudi Emeribe, Lawrence Ezemonye and Mike Ajieh

Abstract

Amid global challenges like climate change and resource depletion, Nigeria's pursuit of low-carbon energy resources teeters on the brink. Nigeria grapples with the delicate balance between economic growth, environmental stewardship, and energy security. Energy and natural resources are crucial to modern society, playing a pivotal role in economic development, environmental sustainability, and overall human well-being. In response to these challenges, the national energy research centers have emerged as vital institutions that can contribute to the development of sustainable innovations that would catalyze the transition to a low-carbon economy. Unlocking these potential hinges on cutting-edge research tailored to drive innovation, influence policy, and promote responsible resource management. Investing in research holds the potential to facilitate Nigeria's transition to a low-carbon economy. These research centers can act as catalysts for unleashing the full potential of Nigeria's energy and natural resources contributing to global efforts to combat climate change while ensuring economic growth and energy security for the populations. Achieving these objectives requires an increased investment in avant-garde research, which would propel Nigeria toward a future characterized by sustainability and prosperity and paving the way for a resilient and environmentally conscious path to long-term development.

Keywords: energy research centers, natural resources, energy security, climate change, renewable energy

1. Introduction

The exponential growth in the world's population places strain on natural resources demand. Major countries around the world are strengthening their resource strategies to guarantee long-term stable supply. With a rise from 30 billion tons in

1970 to 70 billion tons in 2010, the world's yearly material extraction has increased dramatically [1]. Global energy markets are experiencing significant transformations, with growing demand from emerging economies, increasing awareness of environmental considerations, and advances in new technologies. Energy and natural resources are essential pillars of the global economy, exerting a profound impact on the development of the world's energy landscape. They encompass various materials, components, and equipment utilized in the production, conversion, and distribution of renewable, and conventional energy sources. As key elements of the global economy, they play a pivotal role in shaping the trajectory of energy development worldwide.

Africa possesses abundant natural resources, including oil, gas, coal, minerals, and renewable energy sources like solar, wind, hydroelectric, geothermal, and biomass yet faces severe energy deficiencies, with a large portion of its population lacking access to modern energy services [2–5]. This paradox emphasizes a critical challenge facing the continent, where abundant resources coexist with pervasive energy poverty. Energy poverty not only impacts the daily lives of millions but also impedes the region's potential for economic growth and social advancement. It stands as a critical indicator closely tied to the well-being, health, gender equality, poverty levels, and food security of households and nations [2, 6–12].

Africa's excessive reliance on fossil fuels has perpetuated the cycle of energy poverty, characterized by inadequate access to electricity, clean fuel, and modern energy infrastructure. Millions of Africans lack access to reliable electricity, relying instead on inefficient and environmentally harmful energy sources such as biomass, kerosene, and diesel generators [13]. Within the framework of the United Nations' Sustainable Development Goals (SDGs), several goals specifically address energy and natural resources, underscoring the importance of promoting sustainable practices and ensuring access to clean energy for all. SDG 7 (Affordable and Clean Energy), SDG 9 (Industry, Innovation, and Infrastructure), SDG 11 (Sustainable Cities and Communities), SDG 12 (Responsible Consumption and Production), SDG 13 (Climate Action), SDG 14 (Life Below Water), and SDG 15 (Life on Land) collectively aim to advance sustainable energy and natural resource management practices. Energy poverty (SDG 7) and greenhouse gas emissions (SDG 13) represent significant hurdles on the path toward achieving sustainable development goals in Africa. The dual challenges of energy poverty and climate change necessitate urgent action to transition toward cleaner, more sustainable energy systems. By investing in renewable energy infrastructure, improving energy efficiency, and promoting inclusive energy access, African nations can unlock opportunities for economic growth, environmental conservation, and social development.

In the heart of Africa lies Nigeria, a nation richly endowed with a plethora of energy resources, ranging from conventional non-renewable sources to vast reserves of renewable energy like solar, wind, hydroelectric, geothermal, and biomass [14, 15] (**Tables 1–3**). Despite this wealth of resources, Nigeria continues to grapple with the scourge of energy poverty, with a significant portion of its population deprived of access to affordable, reliable, and modern energy services. This dissonance between resource abundance and energy prosperity accentuates the complexities inherent in addressing energy poverty. Nigeria's energy landscape presents a paradoxical narrative. On the one hand, the country boasts substantial reserves of oil and natural gas, making it a prominent player in the global energy market. On the other hand, the potential offered by renewable energy sources remains largely untapped, despite their abundance and suitability for sustainable development. As a result, millions of

S/No	Resource type	Reserves (natural units)		Product level (natural units)	Utilization (natural units)
1	Light crude oil	37.06 billion bbl (as of 2017)		2.5 million barrels/day (as of 2014)	316,000 bbl/day (as at 2015)
2	Natural gas	5.284 trillion cubic meter (CUM) (as of 2017)		45.15 billion CUM (as of 2014)	26.86 billion CUM (as of 2015)
3	Coal and lignite	2.734 billion tonnes		Insignificant	Insignificant
4	Tar sands	31 billion barrels of oil equivalent		—	—
5	Large hydropower	11,250 MW		1938 MW (167.4 million MWh/day)	167.4 million MWh/day
6	Small hydropower	3500 MW		30 MW (2.6 million MWh/day)	2.6 million MWh/day
7	Solar radiation	3.5–7.0 kWh/m^2/day (485.1 million MWh/day using 0.1% Nigeria land area)		Excess of 240 kWp of solar PV or 0.01 million MWh/day	Excess of 0.01 million MWph/day of solar PV
8	Wind	2–4 m/s at 10 m height		—	—
9	Biomass	Fuelwood	11 million hectares of forest and woodland	0.110 million tonnes/day	0.120 million tonnes/day
		Animal waste	245 million assorted in 2001	0.781 million tonnes of waste/day	245 million assorted in 2001
		Energy drops and agricultural residues	72 million hectares of agricultural land and all wastelands	72 million hectares of agricultural land and all wastelands	72 million hectares of agricultural land and all wastelands
10	Nuclear element		Not yet quantified	—	—

Table 1.
Energy resources of Nigeria.

S/No	Region	State	Research center	Host university	Year founded	Research role
1	North-East (NE)	Adamawa Bauchi Borno Gombe Taraba Yobe	National Center for Petroleum Research and Development (NCPRD)	Abubakar Tafawa Balewa University Bauchi, Bauchi State	2008	Petroleum
2	North-West (NW)	Jigawa Kaduna Kano Katsina Kebbi Sokoto Zamfara	Sokoto Energy Research Center (SERC)	Usman Danfodiyo University Sokoto, Sokoto State	1982	Solar and renewable energy

S/No	Region	State	Research center	Host university	Year founded	Research role
3	South-South (SS)	Akwa Ibom Bayelsa Cross River Rivers Delta Edo	National Center for Energy and Environment (NCEE)	University of Benin, Edo State	2009	Energy and environment
4	North-Central (NC)	Benue Kogi Kwara Nasarawa Niger Plateau FCT	National Center for Hydropower Research and Development (NCHRD)	University of Ilorin, Ilorin, Kwara State	2008	Hydropower
5	South East (SE)	Abia Anambra Ebonyi Enugu Imo	The National Center for Energy Research and Development (NCERD)	University of Nigeria, Nsukka, Anambra State	1980	Solar and renewable energy
6	South-West (SW)	Ekiti Lagos Ogun Ondo Osun Oyo	National Center for Energy Efficiency and Conservation (NCEEC)	University of Lagos, Lagos State	2008	Energy efficiency and conservation

Table 2.
Regional distribution of energy research centers in Nigeria.

Resource type	Reserves	Reserve billion metric tons of oil equivalent (BToE)
Crude oil	36.2 billion barrels	4896
Natural gas	166 trillion standard cubic feet (SCF)	4465
Coal and lignite	2.7 billion tonnes	1882
Tar sands	31 billion barrels of oil equivalent	4216
Subtotal fossil	15,459	
Hydropower large scale	11,000 MW	
Hydropower small scale	3250 MW	
Fuelwood	13,071,464 ha	
Animal waste	61 million tonnes/year	
Crop residue	83 million tones/year	
Solar radiation	3.5–7.0 kW h/m^2/day	
Wind	2–4 m/s annual average at height 10 m	

Table 3.
Nigeria's energy reserve/capacity.

Nigerians continue to endure the hardships of inadequate energy access, hindering socio-economic progress and perpetuating cycles of poverty.

According to the World Bank, nearly 45% of Nigeria's population still struggles with the darkness of energy deprivation, equating to approximately 99 million individuals living without access to electricity [16]. This staggering statistic underscores the urgent need for comprehensive strategies and concerted efforts to bridge the gap between energy availability and accessibility, ensuring that Nigeria's energy wealth translates into tangible benefits for its citizens. The urgent need for a transition to cleaner energy sources offers a dual opportunity for Nigeria [7]. Firstly, it presents a chance to mitigate the adverse environmental impacts associated with fossil fuel dependency, including air pollution and climate change. Secondly, it unlocks potential economic benefits through enhanced energy access (**Table 2**) and the adoption of modular renewable energy technologies, fostering economic growth and social development. Despite facing numerous challenges within its energy sector, including inadequate electricity generation capacity and unreliable power supply, Nigeria stands at a pivotal moment in its energy transition journey. By prioritizing renewable energy investments, implementing supportive policies, and fostering innovation, Nigeria can pave the way toward a sustainable and resilient energy future.

2. Energy and natural resource potential in Nigeria

Nigeria boasts significant energy and natural resource potential, including vast reserves of oil, natural gas, coal, and renewable energy sources such as solar and wind (**Table 1**) [9, 10]. However, challenges such as energy poverty, inadequate infrastructure, and environmental degradation accentuate the need for sustainable development strategies. With substantial proven oil reserves estimated at over 37 billion barrels, Nigeria ranks among the top oil-producing nations globally. The Niger Delta region, particularly, has been a focal point of oil exploration and production, driving significant revenue for the country. Additionally, Nigeria's vast natural gas reserves have positioned it as a key player in the global liquefied natural gas (LNG) market. Leveraging these resources efficiently presents an avenue for economic diversification and energy security.

Beyond oil and gas, Nigeria has significant reserves of coal, lithium, uranium, cobalt, and rare earth elements (REEs). Coal, historically utilized for electricity generation and industrial processes, faces environmental challenges but is still utilized to a limited extent for energy production [13]. Lithium-ion batteries have become increasingly important for storing energy from renewable sources like solar and wind power. Lithium, a lightweight metal, is a key component of these batteries due to its high energy density and rechargeability. Demand for lithium has surged with the growth of electric vehicles and grid-scale energy storage systems. Uranium is a radioactive element used as fuel in nuclear power plants to produce electricity. Nuclear energy is a low-carbon alternative to fossil fuels, emitting negligible greenhouse gases during electricity generation. Cobalt is another critical component of lithium-ion batteries, serving to stabilize their performance and increase energy density. REEs in the Southwestern and Southeastern of Nigeria are a group of 17 elements crucial for various clean energy technologies, including wind turbines, electric vehicles, and energy-efficient lighting. These minerals are essential for the production of magnets, catalysts, and phosphors used in these applications.

2.1 Carbon emissions and footprint in Nigeria

Despite its rich energy resources, Nigeria stands at a crossroads, grappling with the urgent need to reduce carbon emissions and environmental degradation while harnessing its abundant energy and natural resources to drive sustainable development [17]. Rapid urbanization, industrialization, energy consumption patterns, and deforestation contribute to rising emissions, exacerbating climate change impacts, and threatening public health and ecosystems. As Nigeria continues to develop and industrialize, the issue of carbon emissions and its environmental footprint has become increasingly significant. As of 2022, carbon emissions from the power sector in Nigeria reached around 11.8 million metric tons of carbon dioxide equivalent. The country's energy sector, dominated by fossil fuels such as oil, natural gas, and coal, remains a primary source of carbon emissions. Additionally, deforestation, agricultural practices, and industrial activities contribute to the nation's carbon footprint.

In line with the country's Nationally Determined Contributions (NDC) submitted to the United Nations Framework Convention on Climate Change (UNFCCC), Nigeria has committed to cutting emissions by 20% by 2030, comparable to current levels, with the potential to cut emissions by up to 45% with assistance from international sources [18]. To achieve this, the government has taken several steps to encourage investment in renewable energy, such as creating a feed-in tariff system for renewable energy projects and establishing the Nigerian Renewable Energy and Energy Efficiency Policy (NREEEP) [18]. With the efforts to diversify the energy mix and promote renewable energy sources, the transition remains gradual, resulting in persistent reliance on carbon-intensive fuels. Nigeria recently took on three significant pledges at the Conference of Parties (COP26) in 2021, two of which are to reach net-zero emissions by 2060.

However, to mitigate greenhouse gases (GHGs) and CO_2 emissions, international organizations such as; the United Nations (U.N.), European Union (E.U.), Intergovernmental Panel on Climate Change (IPCC), and Organization for Economic Co-operation and Development (OECD) have suggested carbon tax should be a policy instrument for achieving a given reduction target among numerous administrative reforms and policies including energy transition, environmental related taxes, emission disclosure standards, and emission trading schemes [19]. The carbon tax imposed on CO_2 emissions or carbon content of fossil fuels prevents enterprises from using excess fossil fuels [18]. In spite of the challenges, Nigeria possesses significant opportunities to mitigate its carbon emissions and pursue sustainable development pathways. Transitioning toward renewable energy sources presents a viable solution to reduce reliance on fossil fuels and curb emissions. Effective policy frameworks and institutional mechanisms are essential to drive carbon mitigation efforts in Nigeria. Strengthening regulatory frameworks, enforcing environmental standards, and incentivizing low-carbon investments are crucial steps toward achieving emissions reduction targets.

2.1.1 Nigeria's policies on carbon emissions

Nigeria, like many other countries, has recognized the importance of addressing carbon emissions and reducing carbon footprints to mitigate climate change and promote sustainable development. Various Energy Policies and Initiatives such as Nigeria's National Energy Policy outline the country's energy development goals and strategies. The policy emphasizes the importance of sustainable energy development, including increasing the share of renewable energy sources, improving energy efficiency, and reducing environmental impacts, including carbon emissions. Nigeria is also implementing policies and programs to promote renewable energy development as part of its energy mix diversification and carbon emissions reduction efforts. This includes the National Renewable Energy and Energy Efficiency Policy, which aims to increase the contribution of renewable energy sources such as solar, wind, biomass, and hydropower to Nigeria's energy supply. Nigerian has also launched the Renewable Energy Master Plan (REMP), which seeks to increase the use of renewable energy to 30% by 2030. The plan also aims to increase energy efficiency by 20%.

Improving energy efficiency is a priority in Nigeria's energy policies to reduce carbon emissions and enhance energy security. The Energy Commission of Nigeria and Standard Organization of Nigeria have introduced energy efficiency standards and labeling programs for appliances and equipment, as well as initiatives to promote energy-efficient practices in industries, buildings, transportation, and agriculture. Also, other policies are focusing on expanding natural gas infrastructure and increasing domestic gas utilization for power generation, industrial processes, and transportation to reduce carbon emissions.

While Nigeria has not yet implemented a comprehensive carbon pricing mechanism, discussions around carbon pricing and market-based mechanisms to incentivize emission reductions and promote low-carbon investments have been ongoing. The government has expressed interest in exploring options such as emissions trading schemes and carbon taxation to mitigate carbon emissions. Nigeria is a signatory to international agreements such as the Paris Agreement, which commits countries to take actions to mitigate climate change and reduce greenhouse gas emissions. Nigeria's energy policies align with its commitments under the Paris Agreement, and the country has submitted its Nationally Determined Contributions (NDCs) outlining its targets for emission reductions and climate adaptation measures.

Nigeria is investing in research and development initiatives to support clean energy technologies, carbon capture and storage (CCS), and other innovative solutions to reduce carbon emissions from energy production and consumption. Overall, Nigeria's energy policies aim to promote sustainable energy development, reduce reliance on fossil fuels, and mitigate carbon emissions to address climate change challenges while supporting economic growth and energy security objectives. Continued implementation of these policies, coupled with efforts to strengthen institutional capacity, mobilize investments, and enhance international cooperation, will be crucial for achieving Nigeria's energy and climate goals.

An increasing body of research has highlighted efforts to reduce global energy-related CO_2 emissions in order to reach the carbon neutral (CN) objective [20–22]. Carbon neutrality is attained when the amount of CO_2 emitted into the atmosphere through economic activities is equal to the amount of CO_2 absorbed in carbon sinks such as soil, forest, and ocean through a process known as carbon sequestration [23]. According to Zou et al. [21], carbon neutrality is the point at which human economic interference with global climate systems has negligible discernible consequences.

3. Energy Commission of Nigeria Research Centers

The Energy Commission of Nigeria (ECN), established by Act No. 62 of 1979 and subsequently amended by Act No. 32 of 1988 and Act No. 19 of 1989, plays a pivotal role in strategically planning and coordinating national energy policies across diverse domains. Empowered by its mandate, the ECN functions as the primary governmental body responsible for overseeing energy sector planning and policy implementation. It actively advocates for the diversification of energy sources, promoting the development and efficient utilization of existing resources, while also leading the integration of new and alternative energy options such as Solar, Wind, Biomass, and Nuclear Energy. In addition to its core responsibilities, the Commission serves as a central repository for gathering and disseminating information on national energy policy. It acts as a focal point for resolving technical issues arising from policy implementation and provides advisory services to the Federal and State Governments on energy-related matters upon request. The ECN also develops periodic master plans for the balanced and coordinated advancement of energy in Nigeria, in consultation with relevant government agencies. These plans include recommendations for exploiting new energy sources and other initiatives deemed beneficial to national interests. The Commission establishes guidelines for the utilization of different energy types for specific purposes and investigates funding adequacy in the energy sector, offering recommendations to the government on aspects such as research and development, production, and distribution. It analyzes and publishes pertinent energy information from diverse sources and monitors the sector's performance in alignment with government energy policies.

The ECN collaborates with international organizations such as the International Atomic Energy Agency, International Renewable Energy Agency, and World Energy Council to foster global partnerships and knowledge exchange. It also promotes training and capacity-building initiatives within the energy sector and undertakes any additional tasks as directed by the government. To enhance its coordination (**Figure 1**) and formulation of well-informed policies regarding renewable energy, energy efficiency, and resource management, the ECN established energy research centers to advance knowledge and technologies across diverse energy sectors, encompassing fossil fuels, renewable energy, and energy efficiency (**Table 2**). The ECN meticulously formulates long-term energy plans, strategies, and forecasts to guide Nigeria's energy sector toward sustainability. It systematically collects, analyzes, and disseminates energy data and statistics through its research Centers to inform policy decisions, guide investment planning, and facilitate project implementation nationwide.

Moreover, the ECN assumes a pivotal role in advocating for renewable energy sources, offering incentives, technical support, and capacity-building initiatives to promote the adoption and utilization of renewable energy technologies across Nigeria. Through fostering innovation, collaboration, and knowledge exchange, the ECN remains resolute in propelling Nigeria's energy transition toward a sustainable and prosperous future. Research centers under energy commissions typically focus on various aspects of energy production, distribution, efficiency, and sustainability. The mandates of ECN's energy research centers are tailored to leverage the comparative advantages of the geopolitical regions in which they are situated (**Figure 2**). These mandates are designed to align with the goals and priorities established by the overseeing energy commission.

3.1 National Center for Energy Research and Development (NCERD)

NCERD established in the Southeast (SE) region of Nigeria is one of the flagship research Centers under the Energy Commission of Nigeria. It focuses on conducting research, development, and demonstration activities in various areas of energy, including renewable energy technologies, energy efficiency, and energy policy analysis.

Figure 1.
Coordination of different research centers under ECN.

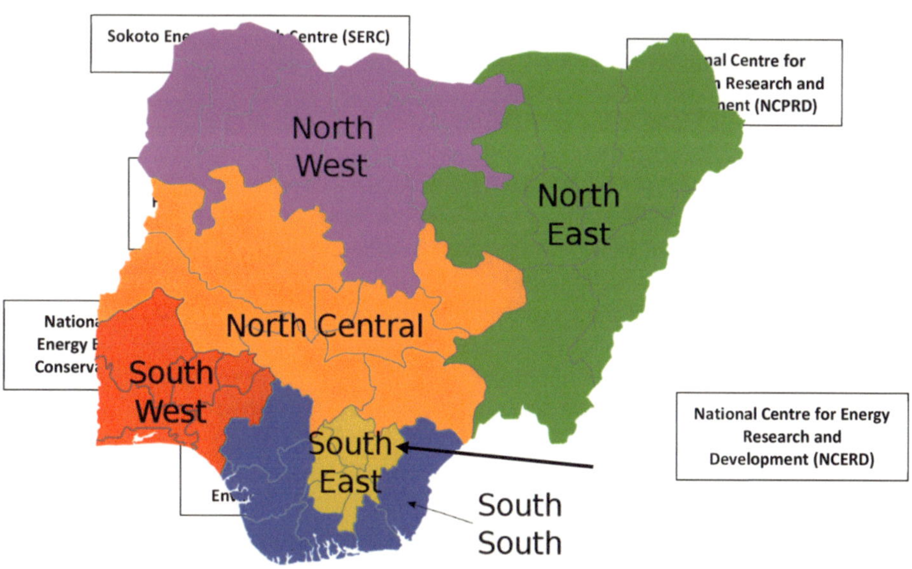

Figure 2.
Distribution of energy research centers in the six geopolitical regions of Nigeria.

3.2 National Center for Hydropower Research and Development (NCHRD)

NCHRD established in the North-Central (NC) region of Nigeria is dedicated to research and development efforts related to hydropower generation in Nigeria. It aims to explore the country's hydropower potential, improve hydropower technologies, and promote sustainable hydropower development practices.

3.3 National Center for Petroleum Research and Development (NCPRD)

NCPRD established in the North-East (NE) focuses on research and development activities in the petroleum sector, including exploration, production, refining, and utilization of petroleum resources. It aims to enhance the efficiency, sustainability, and competitiveness of Nigeria's petroleum industry.

3.4 National Center for Energy Efficiency and Conservation (NCEEC)

NCEEC established in the South-West (SW) region focuses on research and development efforts aimed at promoting energy efficiency and conservation practices across different sectors of the economy. It conducts studies, develops standards, and provides recommendations to improve energy efficiency and reduce energy consumption in Nigeria.

3.5 The Sokoto Energy Research Center (SERC)

SERC established in the North-West (NW) of Nigeria is one of the prominent research Centers under the Energy Commission of Nigeria (ECN). The Center is specifically focused on research and development efforts related to renewable energy and energy efficiency in Nigeria, with a particular emphasis on the North-West region.

3.6 National Center for Energy and Environment (NCEE)

NCEE was established in the South-South (SS) region of Nigeria and is one of the research Centers under the Energy Commission of Nigeria (ECN). The NCEE focuses on research, development, and promotion of sustainable energy and environmental practices in Nigeria. NCEE conducts research on various aspects of energy and the environment, including renewable energy technologies, energy efficiency, climate change mitigation, and environmental sustainability. The Center explores innovative solutions to address energy challenges while minimizing environmental impacts.

4. Collaborative strategies of the energy research centers

Energy Research Centers (ERCs) work together to create a research ecosystem characterized by collaboration, innovation, and impact. By establishing collaborative networks, engaging in joint research projects, sharing resources, providing training and capacity-building opportunities, disseminating knowledge, and engaging in policy advocacy, research centers contribute to the advancement of knowledge, the development of human capital, and the translation of research into meaningful societal outcomes. As Nigeria continues to navigate the complex challenges of energy security and sustainability, the role of energy research centers becomes increasingly crucial. These centers serve as hubs of innovation, driving advancements in technology, policy, and industry practices. Collaboration among these centers can foster innovation, drive technological advancements, and accelerate the transition to renewable energy sources. To achieve visible progress in the research and development programs of the Commission, ERCs must work together to establish a research ecosystem and realize Nigeria's vision of becoming a regional renewable energy powerhouse. This entails employing various strategies to address the nation's energy needs and foster sustainable development. Key strategies include:

4.1 Knowledge sharing and collaboration

Knowledge sharing and collaboration among energy research centers in Nigeria are fundamental pillars for fostering innovation and addressing the nation's energy challenges. By actively exchanging insights, data, and expertise, these centers collectively advance research, drive technological innovations, and accelerate the adoption of sustainable energy solutions. The energy research centers across Nigeria have established platforms for sharing research findings, best practices, and resources. Platforms for knowledge sharing and collaboration include online forums, workshops, seminars, and conferences. These are spaces established for researchers to interact and exchange ideas. These platforms facilitate the dissemination of knowledge and the formation of collaborative networks to address common challenges and leverage diverse expertise. This sharing of knowledge fosters a culture of collaboration and accelerates progress toward renewable energy goals.

4.2 Interdisciplinary research initiatives

Interdisciplinary research initiatives are pivotal for energy research centers in Nigeria to address the multifaceted challenges of the energy sector comprehensively. By integrating expertise from diverse disciplines, such as engineering, environmental science, economics, social sciences, and policy studies, these initiatives generate innovative solutions that are both technically sound and socially relevant. Energy research centers foster interdisciplinary research initiatives by identifying research priorities that require interdisciplinary approaches such as renewable energy integration, energy access in rural communities, sustainable urban planning, and the socioeconomic impacts of energy transitions. By aligning research agendas with societal needs, interdisciplinary initiatives can produce research outcomes with tangible benefits. Also, energy research centers initiate cross-cutting research projects that leverage expertise from different disciplines to address complex energy challenges. For example, a research project on solar energy deployment may involve engineers designing efficient photovoltaic systems, environmental scientists assessing the environmental impacts, economists analyzing the cost-effectiveness, and sociologists studying community acceptance and participation. By integrating diverse perspectives, these projects can produce holistic solutions that account for technical, environmental, economic, and social dimensions. Energy research centers also promote interdisciplinary training and education programs to prepare researchers for collaborative work across disciplines. These programs include interdisciplinary courses, workshops, seminars, and experiential learning opportunities. By fostering interdisciplinary skills such as communication, teamwork, and problem-solving, energy research centers cultivate a new generation of researchers capable of addressing complex energy challenges in a holistic manner. These initiatives can provide comprehensive solutions to complex energy issues and facilitate the integration of renewable energy technologies into Nigeria's energy landscape.

4.3 Joint funding and resource mobilization

Pooling resources through joint funding mechanisms enhances the capacity of energy research centers to undertake ambitious projects. Collaborative grant applications and partnerships with government agencies, industry stakeholders, and international organizations provide the necessary funding and support for impactful research endeavors. ERCs form consortia and networks with other research institutions, universities, and non-governmental organizations (NGOs) to access funding opportunities and share resources. Collaborative research consortia strengthen grant applications, enhance competitiveness, and enable resource sharing among partner institutions. Networks facilitate knowledge exchange, collaboration, and coordination of research efforts across different organizations. Energy commission research centers (ECRCs) establish strategic alliances with academic institutions, research laboratories, and industry associations through formal agreements such as memorandum of understandings (MoUs) to facilitate joint funding and resource mobilization efforts. These alliances enable collaborative research, technology transfer, and capacity-building initiatives that benefit all parties involved.

4.4 Infrastructure sharing and capacity building

Sharing research infrastructure and facilities optimizes resource utilization and reduces duplication of efforts. Additionally, collaborative training programs and workshops can enhance the skills and expertise of researchers, technicians, and students across different centers, thereby strengthening the research ecosystem.

4.5 Stakeholder engagement and policy advocacy

ERCs collectively engage with policymakers, industry leaders, and civil society organizations to advocate for supportive policies and regulations conducive to renewable energy development. By providing evidence-based recommendations and expert insights, research centers influence decision-making processes and drive the adoption of renewable energy technologies.

5. Recent projects by Nigeria's energy research centers

Through various initiatives and collaborations, the ECN's research centers have made significant strides in harnessing clean energy resources, promoting sustainable livelihoods, and remedying environmental issues. One of the notable projects undertaken by the ECN's research centers is the Professional Training Courses on Solar Photovoltaic Installation (SPVI). In collaboration with partners such as the European Union (EU), Deutsche Gesellschaft für Internationale Zusammenarbeit (GIZ), and United States Agency for International Development (USAID), the centers provided comprehensive training to equip individuals with the skills needed to install and maintain solar photovoltaic systems. Furthermore, special attention was given to empowering women and youth through training programs focused on renewable energy technologies, thus fostering inclusivity and diversity within the sector.

In addition to SPVI, the ECN's research centers conducted training sessions on Solar Thermal Systems for Experts and Professionals, in collaboration with the Economic Community of West African States (ECOWAS). These programs aim to enhance the expertise and proficiency of professionals in utilizing solar thermal systems, thereby contributing to the widespread adoption of sustainable energy solutions across various sectors.

Recognizing the importance of energy efficiency in mitigating environmental impact and reducing energy consumption, the ECN's research centers have undertaken several initiatives in this area. Through the Training-of-Trainers (ToT) program on Energy Audit and Energy Management, funded by the German Government under the Nigerian Energy Support Program (NESP), professionals were equipped with the necessary knowledge and skills to conduct energy audits and implement effective energy management practices. By promoting energy efficiency measures, these initiatives not only contribute to cost savings but also help in reducing carbon emissions and enhancing sustainability. Furthermore, the centers have spearheaded activities aimed at retrofitting incandescent lamps with compact fluorescent lamps (CFL). This simple yet effective measure not only helps in reducing energy consumption but also serves as a practical demonstration of energy-efficient lighting solutions. By promoting the adoption of CFLs, the ECN's research centers play a vital role in advocating for sustainable practices at the grassroots level.

In line with their commitment to environmental sustainability, the ECN's research centers have developed biogas technologies as a renewable energy source. By utilizing organic waste materials, such as agricultural residues and animal manure, these technologies not only generate clean energy but also contribute to waste management and environmental conservation. Moreover, the centers have been actively involved in environmental forensics and remediation for energy-related activities. By conducting thorough assessments and implementing remedial measures, they address environmental concerns arising from energy utilization practices, thereby promoting responsible and sustainable energy development. The Energy Commission of Nigeria's research centers have been instrumental in driving forward the agenda of renewable energy, energy efficiency, and environmental sustainability. Through their pioneering projects and collaborative efforts, they have not only advanced technological innovations but also empowered individuals and communities to embrace clean energy solutions. As Nigeria strives toward a greener and more sustainable future, the ECN's research centers continue to play a pivotal role in shaping the country's energy landscape for the better.

6. Role of energy research centers in reducing carbon footprint

In Nigeria where energy demands are high and natural resources are abundant reducing the carbon emissions and footprint across various sectors is crucial for achieving sustainable development goals [24]. ERCs play a pivotal role in research and development efforts to reduce energy-related CO_2 emissions in order to reach the carbon neutral (CN) objective. Carbon neutrality is attained when the amount of CO_2 emitted into the atmosphere through economic activities is equal to the amount of CO_2 absorbed in carbon sinks such as soil, forest, and ocean through a process known as carbon sequestration [23]. According to Zou et al. [21], carbon neutrality is the point at which human economic interference with climate systems has negligible discernible consequences. Carbon conversion, emission reduction, and transfer are some of the strategies targeted at achieving carbon neutrality, although many of them are still poorly defined or executed in Nigeria. ERCs' contributions to achieving carbon neutrality can be categorized into several key areas:

6.1 Advancing clean energy technologies

At the heart of ERC's mission lies the development of clean and renewable energy technologies. ERCs invest significant resources into exploring and refining alternative sources of energy such as solar, wind, hydroelectric, and geothermal power. as well as energy storage solutions like batteries and hydrogen fuel cells. ERCs conduct fundamental and applied research to improve efficiency, affordability, and scalability. ERCs are at the forefront of developing clean energy technologies tailored to the country's specific needs and resources. By harnessing the power of innovation, they pave the way for a future where carbon-intensive fossil fuels are gradually phased out in favor of cleaner alternatives. By investing in indigenous clean energy solutions, these centers help reduce Nigeria's reliance on fossil fuels and lower its carbon emissions. ERCs often serve as testbeds for demonstrating the feasibility and effectiveness of emerging clean energy technologies at scale. By showcasing real-world applications, they help bridge the gap between laboratory research and commercial deployment, thereby accelerating the adoption of clean energy solutions.

6.2 Enhancing energy efficiency

Improving energy efficiency is a key strategy for reducing carbon footprint, and ERCs play a vital role in this endeavor. Through research and development initiatives, they identify opportunities to optimize energy consumption in industries, transportation, and buildings. By promoting the adoption of energy-efficient technologies and practices, they contribute to lowering energy demand and mitigating greenhouse gas emissions. This includes designing more efficient appliances, and industrial processes, as well as implementing smart grid technologies and energy management systems. By reducing energy waste and maximizing the use of resources, energy research Centers contribute significantly to lowering carbon emissions and mitigating the impacts of climate change. In addition to developing new sources of clean energy, energy research Centers are deeply committed to improving energy efficiency across all sectors of the economy. Through rigorous analysis and experimentation, researchers identify opportunities to optimize energy consumption in industries, transportation, and buildings. This includes designing more efficient appliances, vehicles, and industrial processes, as well as implementing smart grid technologies and energy management systems.

6.3 Promoting renewable energy integration

ERCs collaborate with stakeholders to promote the seamless integration of renewable energy into Nigeria's energy mix. They work on grid modernization initiatives, energy storage solutions, and demand-side management strategies to facilitate the integration of solar, wind, and other renewable sources into the existing energy infrastructure. By bolstering the reliability and stability of renewable energy systems, the Centers accelerate the transition toward a low-carbon energy future. However, the deployment of renewable energy technologies remains relatively low, constrained by challenges such as policy inconsistency, limited financing options, and technical barriers. Promoting renewable energy integration is essential for reducing carbon footprints and transitioning toward a more sustainable energy future.

6.4 Advancing carbon capture and storage (CCS)

Given Nigeria's significant role as an oil-producing nation, ERCs focus on developing carbon capture and storage (CCS) technologies to mitigate emissions from fossil fuel-based industries. These Centers conduct research on cost-effective CO_2 capture methods and explore suitable storage sites, laying the groundwork for the deployment of CCS projects across the country. By capturing and sequestering carbon emissions, they help Nigeria meet its climate goals while sustaining its energy-intensive industries. They also investigate safe and effective ways to store captured carbon underground, preventing it from entering the atmosphere and exacerbating global warming. By advancing CCS solutions, energy research centers offer a viable pathway for reducing the carbon footprint of fossil fuel-based energy generation while facilitating a smoother transition to cleaner energy sources.

6.5 Policy development and advocacy

ERCs collaborate with the government and other stakeholders to develop and advocate for policies that promote clean energy adoption and carbon reduction initiatives. They provide evidence-based recommendations, conduct policy analysis, and facilitate stakeholder dialogs to inform the design and implementation of effective energy policies. By driving policy change and creating an enabling regulatory environment, ERCs support Nigeria's transition toward a sustainable, low-carbon energy sector. ERCs advocate for renewable energy incentives, energy efficiency standards, carbon pricing mechanisms, and other policy measures to promote sustainable energy development, reduce greenhouse gas emissions, and mitigate climate change impacts.

ERCs have been instrumental in advocating for renewable energy deployment and supporting policy reforms to create an enabling environment for sustainable energy development in Nigeria. Through policy analysis, research reports, and stakeholder engagement, ERCs have provided evidence-based recommendations to policymakers, legislators, and regulators on the need for renewable energy incentives, feed-in tariffs, tax incentives, and regulatory reforms. This advocacy has led to the development and implementation of supportive policies and programs, such as the National Renewable Energy Action Plan (NREAP), Renewable Energy Master Plan (REMP), and Renewable Energy Policy (RE Policy), which aim to promote the deployment of renewable energy technologies and attract investment in the sector. ERCs have also collaborated with government agencies, industry associations, and civil society organizations to raise awareness about the socio-economic and environmental benefits of renewable energy and build consensus for policy action.

6.6 Capacity building and knowledge sharing

ERCs play a crucial role in building local capacity and fostering knowledge sharing in the energy sector. Through training programs, workshops, and technical assistance initiatives, they empower researchers, policymakers, and industry professionals with the skills and expertise needed to drive innovation and sustainable development.

By nurturing a culture of learning and collaboration, these Centers contribute to long-term solutions for reducing Nigeria's carbon footprint and building resilience to climate change.

ERCs play a vital role in education and capacity building by providing training programs, workshops, and seminars to build the skills and knowledge of energy professionals, policymakers, and stakeholders. They offer professional development programs, and hands-on training in renewable energy technologies, energy management practices, and sustainability principles. ERCs also support research scholarships, fellowships, and internships to cultivate the next generation of energy leaders and innovators, fostering a skilled workforce capable of driving sustainable energy transitions.

ERCs raise public awareness and promote knowledge dissemination on sustainable energy issues through outreach programs, public lectures, and media campaigns. They communicate scientific findings, policy insights, and best practices to inform and empower communities, decision-makers, and stakeholders to take action on energy sustainability. ERCs engage in community outreach, environmental education, and public engagement initiatives to foster a culture of energy conservation, environmental stewardship, and sustainable development at the local, national, and global levels.

Overall, ERCs play a multifaceted role in advancing sustainable energy development by driving innovation, informing policy, building capacity, fostering collaboration, and raising awareness. Their contributions are essential for achieving the goals of energy security, environmental protection, and socio-economic prosperity in a rapidly changing energy landscape. By harnessing the collective expertise and resources of ERCs, societies can accelerate the transition to a sustainable energy future and address the pressing challenges of climate change, energy poverty, and resource scarcity facing the world today.

6.7 Technology transfer and commercialization

ERCs facilitate technology transfer and commercialization by translating research findings into practical applications, products, and services for the market. They collaborate with industry partners, startups, and entrepreneurs to pilot, scale, and commercialize innovative energy technologies and solutions. ERCs provide technical assistance, incubation support, and access to funding and resources to accelerate the development and deployment of clean energy technologies, fostering economic growth, job creation, and market competitiveness in the energy sector.

6.8 Specialized training programs and courses

ERCs organize specialized training programs, workshops, and courses tailored to various aspects of energy research, technology, and management. These programs cover topics such as renewable energy technologies, energy efficiency measures, energy policy and regulation, sustainable energy systems, and climate change mitigation strategies. Participants, including students, researchers, professionals, policymakers, and industry stakeholders, benefit from hands-on learning experiences, expert lectures, and interactive sessions conducted by leading experts in the field. The training programs are designed to enhance participants' technical skills, analytical capabilities, and problem-solving abilities, preparing them for careers in the energy sector and related industries.

6.9 Access to state-of-the-art laboratories and research facilities

ERCs provide access to state-of-the-art laboratories, research facilities, and experimental infrastructure to support hands-on learning, experimentation, and technology development. These facilities are equipped with advanced instrumentation, testing equipment, computational tools, and simulation software necessary for conducting experiments, analyzing data, and validating research findings. Students, researchers, and industry collaborators have the opportunity to work in interdisciplinary teams, collaborate on research projects, and leverage cutting-edge technologies to address real-world energy challenges. Access to research facilities fosters innovation, collaboration, and knowledge exchange among stakeholders, leading to breakthroughs in renewable energy, energy efficiency, and clean technologies.

Capacity building and training programs offered by ERCs are essential for developing a skilled workforce, fostering innovation, and advancing knowledge in the energy sector. By investing in human capital development, ERCs contribute to building a sustainable energy future and addressing the pressing global challenges of energy security, environmental sustainability, and climate change mitigation. These programs empower individuals to become leaders, innovators, and change agents in shaping the transition to a cleaner, more resilient energy system. The impact of Energy Research Centers (ERCs) in Nigeria is profound, with their initiatives contributing significantly to sustainable energy development across various sectors.

7. Nigeria's energy transition unlocking energy and natural resources for sustainable development

With an increasing global focus on sustainability and climate change mitigation, Nigeria's energy transition is gaining prominence [9]. Energy transition is driven by a combination of factors, including policy frameworks, technological advancements, and economic imperatives. Nigeria has implemented various policy frameworks to promote renewable energy adoption and energy efficiency. The National Renewable Energy and Energy Efficiency Policy (NREEEP), launched in 2015, sets ambitious targets for increasing the share of renewable energy in the energy mix. Additionally, initiatives such as the Rural Electrification Agency's Mini-Grids Program aim to enhance energy access in rural areas through decentralized renewable energy solutions. The deployment of renewable energy technologies, particularly solar and hydroelectric power, has witnessed significant progress in Nigeria. Large-scale solar projects, such as the Katsina Solar Power Plant, highlight the country's potential to harness solar energy resources.

Nigeria's energy transition is a complex and multifaceted process that requires concerted efforts from policymakers, industry stakeholders, and civil society. While significant progress has been made, addressing existing challenges and capitalizing on emerging opportunities are crucial for advancing toward a sustainable energy future. By leveraging its abundant renewable energy resources and fostering an enabling environment for investment and innovation, Nigeria can navigate its energy transition path and contribute to global efforts to mitigate climate change while ensuring energy security and socioeconomic development. In parallel with efforts to reduce carbon emissions, research Centers focus on unlocking Nigeria's energy and natural resource potential for sustainable development.

8. Challenges and opportunities

Despite progress, the Research Center's efforts to tackle carbon emissions and unlock energy resources face challenges such as funding constraints, policy barriers, and institutional capacity gaps. Weak governance structures, regulatory inconsistencies, and corruption hinder effective management and equitable distribution of resource revenues. Lack of transparency and accountability in the allocation of licenses and contracts further exacerbates socio-economic inequalities and undermines public trust.

Nigeria's energy sector is vulnerable to security risks, including militancy, pipeline vandalism, and insurgency activities in oil-producing regions. These challenges disrupt operations, deter investment, and undermine the stability of the energy market, hampering economic growth and development. The adverse effects of climate change, such as rising temperatures, erratic rainfall patterns, and extreme weather events, pose significant risks to Nigeria's energy and natural resource sector. Increased frequency of droughts, floods, and storms threatens infrastructure, agriculture, and livelihoods, necessitating adaptation and resilience measures.

9. Conclusion

In conclusion, research Centers play a crucial role in addressing carbon emissions and unlocking energy resources for sustainable development in Nigeria. By driving innovation, shaping policy, and fostering collaboration, research Centers contribute to a greener, more resilient future for Nigeria. Moving forward, sustained investment, policy support, and cross-sectoral cooperation are essential to realizing Nigeria's potential as a leader in sustainable energy and natural resource management. ERCs serve as catalysts for advancing clean energy technologies and reducing carbon footprints by conducting R&D, demonstrating technology viability, fostering collaborative partnerships, supporting policy development, educating stakeholders, conducting techno-economic analysis, and nurturing innovation ecosystems. Their multidisciplinary approach and cross-sectoral collaborations are essential for addressing the urgent challenges of climate change and sustainable development.

Conflict of interest

The authors declare no conflict of interest.

DOI: http://dx.doi.org/10.5772/intechopen.1005050

Author details

Emmanuel Ogbomida[1,2,3]*, Abdullahi Mustapha[4], Chukwudi Emeribe[2], Lawrence Ezemonye[5] and Mike Ajieh[6]

1 Copperbelt University, Kitwe, Zambia

2 National Center for Energy and Environment, (Energy Commission of Nigeria), University of Benin, Benin City, Edo State, Nigeria

3 Laboratory of Toxicology, Department of Environmental Veterinary Sciences, Graduate School of Veterinary Medicine, Hokkaido University, Sapporo, Japan

4 Energy Commission of Nigeria, Abuja, Nigeria

5 Department of Animal and Environmental Biology, University of Benin, Benin City, Edo State, Nigeria

6 Department of Chemical Engineering, Delta State University, Abraka, Oleh, Nigeria

*Address all correspondence to: ogbomida.e@ncee.org.ng

References

[1] Bai Z, Ma W, Zhao H, Guo M, Oenema O, Smith P, et al. Food and feed trade have greatly impacted global land and nitrogen use efficiencies over. Nature Food. 1961-2017;**2**(10):780-791

[2] International Renewable Energy Agency (IRENA), African Development Bank (AfDB). Renewable Energy Market Analysis: Africa and its Regions. Abu Dhabi and Abidjan: IRENA and AfDB; 2022. Available from: https://www.irena.org/-/media/Files/IRENA/Agency/Publication/2022/Jan/IRENA_Market_Africa_2022.pdf?rev=bb73e285a0974bc996a1f942635ca556

[3] Ochola WO, Sanginga PC, Bekalo I. Managing Natural Resources for Development in Africa: A Resource Book. Ottawa: IDRC; 2010

[4] African Capacity Building Foundation (ACBF). African Capacity Indicators 2013 - Capacity Development for Natural Resource Management. ACBF: Harare; 2013

[5] Watkins K. Solar Power to the People. How the Sun Can Ease Africa's Electricity Crisis. The Guardian; 5 June 2015. Available from: https://www.theguardian.com/business/economics-blog/2015/jun/05/solar-power-africa-sun-electricity-crisis

[6] Bilgili F, Ozturk I, Kocak E, Kuskaya S, Cingoz A. The nexus between access to electricity and CO_2 damage in Asian Countries: The evidence from quantile regression models. Energy and Buildings;**256**:111761. DOI: 10.1016/j.enbuild.2021.111761

[7] Chanchangi YN, Adu F, Ghosh A, Sundaram S, Mallick TK. Nigeria's energy review: Focusing on solar energy potential and penetration. Environment, Development and Sustainability;**25**(7):5755-5796. DOI: 10.1007/s10668-022-02308-4

[8] Day R, Walker G, Simcock N, et al. Energy Policy;**93**:255-264. DOI: 10.1016/j.enpol.2016.03.019

[9] Ogbonna CG, Nwachi CC, Okeoma IO, et al. Understanding Nigeria's transition pathway to carbon neutrality using the Multilevel Perspective. Carbon Neutrality. 2023;**2**:24. DOI: 10.1007/s43979-023-00065-5

[10] Oyedepo SO. Energy and sustainable development in Nigeria: The way forward. Energy, Sustainability and Society. 2012;**2**:15. DOI: 10.1186/2192-0567-2-15

[11] Rodriguez-Alvarez A, Orea L, Jamasb T. Fuel poverty and Well-Being: A consumer theory and stochastic frontier approach. Energy Policy. **131**:22-32. DOI: 10.1016/j.enpol.2019.04.031

[12] Wang Q, Kwan M, Fan J, Lin J. Racial disparities in energy poverty in the United States. Renewable and Sustainable Energy Reviews. **137**:110620. DOI: 10.1016/j.rser.2020.110620

[13] Castellano A, Kendall A, Nikomarov M, Swemmer T. Brighter Africa: The Growth Potential of the Sub-Saharan Electricity Sector. McKinsey Report. 2015. 64 p. Available from: http://www.mckinsey.com/industries/electricpower-and-natural-gas/our-insights/powering-africa

[14] Abban OJ, Hongxing Y, Nuta AC, Dankyi AB, Ofori C, Cobbinah J. Renewable energy, economic growth, and CO_2 emissions contained

Co-movement in African oil-producing countries: A wavelet-based analysis. Energy Strategy Reviews. **44**:100977. DOI: 10.1016/j.esr.2022.100977

[15] Amin M, Shah HH, Fareed AG, Khan WU, Chung E, Zia A, et al. Hydrogen production through renewable and non-renewable energy processes and their impact on climate change. International Journal of Hydrogen Energy. 2022;**47**:33112-33134. DOI: 10.1016/j.ijhydene.2022.07.172

[16] World Bank. Access to Electricity (% of Population)—Nigeria | Data (worldbank. org). 2019. [cited 2024 Feb 19]. Available from: https://data.worldbank.org/indicator/EG.ELC.ACCS.ZS?locations=NG

[17] Ahmad F, Draz MU, Chang WY, Yang SC, Su L. More than the resource curse: Exploring the nexus of natural resource abundance and environmental quality in Northwestern China. Resources Policy. 2021;**70**:101902

[18] Li J, Bao-guo F, Rui-peng H, Ben L, Yu-xing Y, Fei H, et al. Study on quenching hydration reaction kinetics and desulfurization characteristics of magnesium slag. Journal of Cleaner Production. 2018;**190**:12-23. DOI: 10.1016/j.jclepro.2018.04.150

[19] Shahzad U. Environmental taxes, energy consumption, and environmental quality: Theoretical survey with policy implications. Environmental Science and Pollution Research. 2020;**27**:24848-24862

[20] Zhao A, Xiaowei M, Boyang C, Yuping S, Malin S. Challenges toward carbon neutrality in China: Strategies and countermeasures. Resources, Conservation and Recycling. 2022;**176**:105-959. DOI: 10.1016/j.resconrec.2021.105959

[21] Zou C, Xue H, Xiong B, Zhang G, Pan S, Jia C, et al. Connotation, innovation, and vision of "carbon neutrality". Natural Gas Industry B. 2021;**8**(5):523-537. DOI: 10.1016/j.ngib.2021.08.009

[22] Relva SG, da Silva VO, Gimenes ALV, Udaeta MEM, Ashworth P, Peyerl D. Enhancing developing countries' transition to a low-carbon electricity sector. Energy. 2021;**220**:119659

[23] European Parliament. What is Carbon Neutrality and How Can It be Achieved by 2050? European Parliament News; 2019. Available from: https://www.europarl.europa.eu/news/en/headlines/society/20190926STO62270/what-is-carbon-neutrality-and-how-can-it-be-achieved-by-2050

[24] Quan Z, Dongmei B, Xiaoxue P. Research status and future trends of natural resources and sustainable development in China: Visual analysis based on CiteSpace. Journal of Resources and Ecology. 2021;**12**:419-429